Please check all items for damages
before leaving the Library.
Thereafter you will be held
responsible for all injuries
to items beyond reasonable wear.

The
Sound
Book

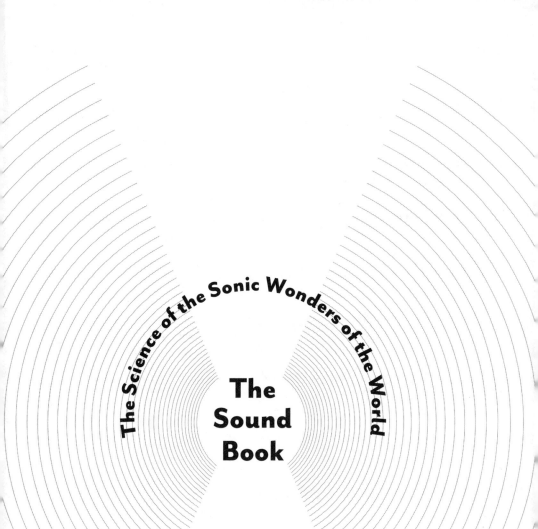

The Science of the Sonic Wonders of the World

The
Sound
Book

Trevor Cox

W. W. Norton & Company

New York / London

Published in Great Britain by The Bodley Head under the title
Sonic Wonderland: A Scientific Odyssey of Sound

Illustration credits: Unless otherwise stated, images are drawn by Trevor and Nathan Cox. Richard Deane took the *Aeolus* photograph (Figure 4.1). The cat piano image (Figure 5.1) is courtesy of CNUM, Conservatoire Numérique des Arts et Métiers, http://cnum.cnam.fr, *La Nature*, 1883, p. 320. The satellite image of clouds showing airflow around Alejandro Selkirk Island (Figure 8.7) is © NASA Goddard Photo and Video's photo stream, http:// www.flickr.com/photos/gsfc/5638320696/in/photostream, accessed January 9, 2013.

For information about permission to reproduce selections from this book,
write to Permissions, W. W. Norton & Company, Inc.,
500 Fifth Avenue, New York, NY 10110

For information about special discounts for bulk purchases, please contact
W. W. Norton Special Sales at specialsales@wwnorton.com or 800-233-4830

Manufacturing by RR Donnelley, Harrisonburg
Book design by Chris Welch
Production manager: Devon Zahn

Library of Congress Cataloging-in-Publication Data

Cox, Trevor J.
The sound book : the science of the sonic wonders of the world /
Trevor Cox. — First American edition.
pages cm
Includes bibliographical references and index.
ISBN 978-0-393-23979-9 (hardcover)
1. Sounds—Popular works. 2. Noise—Popular works. I. Title.
QC225.3.C69 2014
550.1'534—dc23
2013034491

W. W. Norton & Company, Inc.
500 Fifth Avenue, New York, N.Y. 10110
www.wwnorton.com

W. W. Norton & Company Ltd.
Castle House, 75/76 Wells Street, London W1T 3QT

1 2 3 4 5 6 7 8 9 0

To Deborah

Contents

THE SONIC WONDERS OF THE WORLD

Bubbling mud pot
ICELAND

World's most reverberant spa
INCHINDOWN

Silent anec
chambe
SALFORD

Echo Bridge
MASSACHUSETTS

Singing sands of Kelso Dunes
MOJAVE DESERT

Great Stalacpipe Organ
VIRGINIA

Musical road
CALIFORNIA

Chirping Mayan pyramid
MEXICO

Echolocating oilbirds
VENEZUELA

"Mighty noise" tidal bore
BRAZIL

0	1,000	2,000	3,000	4,000 mi
0		2,000	4,000	6,000 km

Bearded seals singing glissandos
SVALBARD

Ice melting on Lake Baikal
SIBERIA
●

Derelict radome at Teufelsberg
BERLIN
●

Ancient theater Epidaurus
GREECE
●

Echo in Imam Mosque
ISFAHAN
●

King Seongdeok Divine Bell
SOUTH KOREA
●

Whispering gallery in Gol Gumbaz
INDIA
●

Rock gongs
SERENGETI NATIONAL PARK
●

Superb lyrebirds
AUSTRALIA
●

The
Sound
Book

Prologue

"Is it safe?" A noxious odor was invading my nostrils as I stared down the open manhole. The metal ladder disappeared into the darkness. I had assumed a radio interview on the acoustics of sewers would involve an official and authorized visit. Instead, it started with a walk into a London park on a summer's evening. Bruno, the interviewer, produced a large key from his knapsack, opened up a convenient manhole cover, and invited me to climb down. Was it legal to wander around the sewers without permission? What if the tunnel suddenly flooded? What about a canary to warn of poisonous gases? Meanwhile, strolling commuters ignored us as we gazed into the gloom.

I repressed my anxieties and climbed gingerly down the ladder to the sewer about 6 meters (20 feet) below. This was a storm drain built in Victorian times, a long cylindrical tunnel lined with bricks. The floor was treacherous and slippery, and the odor made my skin crawl. I clapped my hands as best I could with rubber gloves on and started to count in my head very slowly—"one, two, three, . . ."—timing how long it took the sound to die away. After 9 seconds a distant rumbling echo returned to me. Sound travels a kilometer (about a half mile) every 3 seconds, so my clap had traveled a round-trip of 3 kilometers (1.9 miles). Later on, far away down the tunnel, we discovered the staircase off which the sound had bounced; it was draped in disgusting debris.

I found it difficult to avoid head-butting the stalactites hanging from the low ceiling. Sadly, these were not brittle rock, but crusty, fatty deposits clinging to the bricks. These foul stalactites broke off, worked their way down the back of my shirt and scraped my skin. Since I'm tall, my head was very close to the ceiling—the worst place for the revolting stalactites, but the optimal position for observing an unexpected acoustic effect. As the radio interview started, I noticed my voice hugging the walls of the cylindrical tunnel and spiraling into the distance. Speech spun around the inside of the curved sewer like a motorcyclist performing in a Wall of Death. While every other sense was being overwhelmed with revulsion, my ears were savoring a wonderful sonic gem. This impressive spiraling toyed with me as I tried to work out what was causing the effect. It was so different from anything I had experienced before that I started to doubt what I was hearing. Was it just an illusion, with the sight of the cylindrical sewer fooling my brain into thinking the sound was curving? No; when I closed my eyes, the reverberance still embraced my voice and twisted it around the tunnel. What was causing the sound to stay at the edges of the sewer and not cross into the middle? I have worked in architectural acoustics for twenty-five years, yet the sewer contained a sound effect I had not heard before. I also noticed that Bruno's voice was embellished with a metallic twang as it echoed in the sewer. How was that possible in a place devoid of metal? We were surrounded by bricks.

During those hours listening to the sewer, I had an acoustic epiphany. My particular expertise is interior acoustics—that is, the way sound works in a room. Most of my work has focused on discovering ways to mask or minimize unwanted sounds and acoustic effects. Not long after completing my doctorate, I pioneered new ways of shaping room surfaces that now improve the sound in theaters and recording studios around the world. Above the stage of the Kresge Auditorium at the Massachusetts Institute of Technology, you can see the gently undulating reflectors I designed to help musicians hear each other. For a rehearsal hall at the Benslow Music Trust in Hitchin, England, I designed corrugations to adorn a concave wall in order to stop sound reflections from all being focused onto a single point in the room and thereby altering the timbre of the musical instruments.

In recent years I have been researching how poor acoustics and high noise levels in classrooms affect learning. It seems obvious that pupils need to be able to hear the teacher and have a certain amount of quiet to learn, yet there are architects who have designed schools that are acoustic disasters. My bête noire is open-plan schools, where doors and walls are dispensed with, resulting in the noise from one class disturbing others because there is nothing to impede the sound. The Business Academy Bexley in London opened in 2002 and was short-listed for the prestigious Royal Institute of British Architects' Stirling Prize. The open-plan design caused so many noise problems, however, that the school and local education authority had to spend £600,000 ($0.9 million) installing glass partitions.[1] Part of my research into schools involved playing noise at pupils as they tried to complete simple tasks involving reading comprehension or mental arithmetic. In one test, playing the babble of a noisy classroom at a cohort of fourteen- to sixteen-year-olds lowered their cognitive abilities to those of a control group of eleven- to thirteen-year-olds who were working in quieter conditions.

I am currently working with colleagues to improve the quality of user-generated content online. I started the project after getting frustrated listening to distorted and noisy soundtracks on Internet videos. We are developing software that will automatically detect when an audio recording is poor—for instance, checking whether there is wind noise whistling past a microphone. The idea is to alert users to poor sound conditions before they start recording, or to use audio processing to weed out some of the interference, just as a digital camera looks for flaws and automatically adjusts exposure time and focus. But before we can write the software, we are grappling with people's perceptions of audio quality. When you record your child playing in a school concert, does the quality of the recording matter very much? My personal feeling is that audio distortions can be much more important than visual ones. A blurry video with a clear recording of a loved one singing captures that special moment much better than a clear video in which the lyrics are unintelligible and the voice distorted.

But as I splashed about in the sewer, I realized that distortions can sometimes be wonderful. Despite having studied sound intensely for decades, I had been missing something. I had been so busy trying to remove unwanted noise that I had forgotten to listen to the sounds themselves. In the right place a "defect" such as a sound focus, or the metallic, spiraling echo in the sewer, could be fascinating to listen to. Perhaps ugly, strange, and distorted sounds could teach us something about how acoustics works in everyday situations or even how our brain processes sound. By the time I emerged out of the sewer through a manhole in a leafy suburban street, I decided I wanted to find more such unusual acoustic effects. And not just the ugly ones. I wanted to experience the most surprising, unexpected, and sublime sounds—the sonic wonders of the world.

Somewhere on the vast Internet I imagined I would be able to find a list of other strange sounds to experience. But after a lengthy shower scrubbing away the odorous memory of the sewer, and a few hours online, I realized it would not be so easy. The dominance of the visual has in fact dulled all of our other senses, especially our hearing. Our obsession with sight has led us to produce loads of images of bizarre and beautiful places, but surprisingly few recordings of wonderful sounds. Like the Soundkeeper in Norton Juster's classic children's book *The Phantom Tollbooth*, I sense among my fellow citizens a lack of appreciation of subtle sounds and an increase in discordant noises.[2] But rather than lock away sounds and enforce silence as the Soundkeeper does, I wanted to seek out, experience, and celebrate wonderful aural effects. What fascinating sounds are out there if we just "open" our ears? While there are many books on unwanted noise and how to abate it, there are not many on how to listen better—something acoustic ecologists call *ear cleaning*.

> Open [a] book now and gently open the pages and just listen to the sound . . . that's a very complex sound . . . first there is the sound of the thumb or finger as it brushes against the edge of the paper before you turn the page and then there is the sound of the page as it turns.[3]

This is Murray Schafer, the grandfather of acoustic ecology, demonstrating how even a simple object, such as the book in your hands, can make many different sounds. It is "full of possibilities," he writes. This quote from an ear-cleaning exercise comes from a Canadian radio program from the 1970s. No cotton swabs are involved, however; listeners improve their hearing skills by changing how their brains process sound, not by physically cleaning their ears.

Schafer tells his listeners to remove all distractions—"like eating, drinking or smoking: well smoke if you have to but don't let it distract you"—to control your breathing and close your eyes to "amputate the visual sense." The experience could be disconcerting, because although the radio script is reminiscent of a meditation CD, the bossy narration is far from soothing. The recording reminds me of a scene from an old black-and-white espionage film in which a villain is trying to brainwash the hero.

Despite the unnerving tone, the program includes some intriguing exercises: invent an onomatopoeic name for the sound of a hardback book being slammed shut (*thump* or *thud* does not quite work), or predict and then imitate the sound of a piece of paper being scrunched up and thrown against a wall. Nowadays you might have to choose something different to play with—an e-book reader being dropped in the bath?

Schafer is evangelical about ear cleaning, believing that children should do it to improve their sonic sensibilities, and that people who shape our sound world, such as urban planners, should undergo the process regularly. In his seminal book *The Soundscape*, Schafer suggests some other ear-cleaning activities you could try. The technique he uses most often is to get people to declare a moratorium on speaking for a day, while eavesdropping on sounds made by others. He wrote, "It is a challenging and even frightening exercise," and successful participants "speak of it afterward as a special event in their lives."[4] But my colleague and fellow acoustic engineer Bill Davies believes this is taking it too far: "If you want to give people an acoustic epiphany," he told me, "then a short journey on a soundwalk is a better way of going about it."[5]

A soundwalk can be a simple activity. All one has to do is stroll for a couple of hours without saying a word, focusing intently on the sounds of the city or countryside. I first did this with an eclectic group of thirty engineers, artists, and acoustic ecologists. We formed a slow-moving strung-out crocodile weaving through the streets of London. The cacophony of cars, planes, and other people starkly contrasted with our own enforced silence. I felt like an extra in an old B movie, part of a procession of possessed humans being summoned by some alien force—silent zombies walking toward impending doom.

This particular group was re-creating a soundwalk that Murray Schafer and colleagues had first carried out in the 1970s. We followed a set of prescribed exercises—from trying to count the number of propeller aircraft heard flying over the formal gardens in Regent's Park (pointless nowadays, though you can still count jets), to trying to suppress a loud noise by consciously ignoring it. I chose the loudest sound around, a pneumatic drill that was hammering away on Euston Road.

Ignoring a pneumatic drill proved very hard to do; indeed, at first it seemed impossible. Trying to disregard the pounding noise immediately made it more obvious because of the way our hearing works. Seals might be able to close their outer ears when diving, but humans have no way of physically shutting out sound. We have no "earlids," and there is no auditory equivalent of closing our eyes or averting our gaze. Our hearing is constantly picking up sounds. We cannot physically stop the eardrum, the tiny bones in the middle ear, or the tiny hair cells in the inner ear from vibrating. Inevitably, the inner ear generates electrical signals, which travel up the auditory nerve into the brain. Fingernails scraping down a blackboard or the climax to a Beethoven symphony, good sounds or bad—the ear sends the audio upward. The brain then has to work out which sounds are important and must be paid attention to, and which ones can be safely ignored. Something noisy and abrupt, like the roar of a tiger or the squealing of car brakes, catches our attention immediately so that we can fight or take flight. When we hear something less threatening, we have to think and decide which sound we want to attend to.

Auditory attention was first researched after the Second World War, as the military tried to understand why fighter pilots sometimes ignored crucially important audible messages.[6] A typical experiment had people listening on headphones and saying aloud the words they heard through one of their earphones. Simultaneously, researchers played a distracting message in the other ear. After this test, subjects could recall very little about the distracting message. Scientists would make changes to the distracting speech—switching talkers or language, and even playing the sound backward—yet most people failed to spot the changes.[7] Although many of us believe we listen to multiple sounds simultaneously, and even believe that women are better than men at such multitasking, these tests demonstrate that such an ability is an illusion. We listen to one thing at a time and rapidly change our attention from one sound to another.

Consequently, back on Euston Road the only way to quiet the pneumatic drill was to focus very hard on another sound. I used two strangers having a boisterous conversation outside a pub. Trying to actively suppress the drill just made it louder, but switching my attention elsewhere meant I could exploit the brain's amazing cognitive ability to suppress background noise.

During the hours focusing on the soundscape around me, I heard the fleeting melody of birdsong, an unexpected hush in the piazza outside the British Library, an auditory sense of enclosure as I entered the tunnel under Euston Road, and the subtle squelching of an underinflated bicycle tire on pavement. Interesting sounds suddenly became more obvious and audible. I was amazed to hear how different railway stations sounded; the throb of idling diesel trains in King's Cross made that station seem more authentic than St. Pancras or Euston. Of course, it was not all positive; the clatter of cheap rolling suitcases being dragged along platforms and sidewalks proved intensely annoying.

Acoustic ecologists have amazing ears for such sonic subtleties, but soundwalking and ear cleaning can help anyone learn to consciously tune in to such previously overlooked delights. We have at our disposal immense cognitive power to analyze sound—after all, listening to and decoding music and speech is an incredibly complex task—yet it is something we take for granted. A soundwalk reveals that there are sounds in our everyday life that, if we choose to listen to them, will surprise us with their diversity and uniqueness. Even something as mundane as footsteps encompasses a huge range of sounds, from the clack-clack of high heels on marble, to the squeak of sneakers on a gym floor. If we can unconsciously learn to recognize colleagues approaching unseen along a corridor from the rhythm of their walking, what else might we be able to accomplish with dedicated effort? Our ears play an immensely important role in how we perceive the world. In this book I hope to show how we can filter things differently, to move us away from our overreliance on the visual, showing how diverting our attention in this way can enrich our enjoyment and understanding of the spaces we inhabit.

Acoustic ecologists are also concerned with aural conservation. Soundscapes do not need to be preserved in aspic, but we need to make sure that great sounds are not lost through neglect—not just calls from endangered species, but also other sounds that are important to us. Not long after my first soundwalk, I interviewed artists in Hong Kong for a BBC program about endangered sounds. They lamented the loss of the bells in the tower at the Star Ferry pier in Kowloon in 2006, which used to play the Westminster Chimes. Redevelopment and well-meaning renovations can ruin precious acoustic effects, as happened about a century ago in the US Capitol in Washington, DC, when architects altered the dome and dulled the focusing that used to distort the speech of senators. Acoustic scientists and historians have only recently begun documenting, preserving, and reconstructing the acoustics of a very few important places. Combining the latest methods for predicting architectural acoustics, three-dimensional sound reproduction, and new archaeological research, scientists have begun to reveal some of the ancient sounds of Greek theaters and prehistoric stone circles.

Another major threat to the sonic landscape is the smog of transportation noise. Underwater, baleen whales have to sing louder to overcome shipping noise. In cities, birds such as great tits have changed their tunes to be heard above the traffic. Of course, humans suffer as well: Nearly 40 percent of Americans want to change their place of residence because of noise, 80 million EU citizens live in unacceptably noisy areas, and one in three UK citizens have been annoyed by neighbor noise.[8] Unintelligible announcements in train stations, restaurants where you have to shout to hold a conversation, and annoying mobile phone ringtones—we unnecessarily suffer a large number of acoustic deficiencies.

Some of these sound excesses are of our own making. Many of us take a large daily dose of music and speech through headphones that isolate us from the sound of our environments. It has become part of our daily routine: compared with only five years ago, youngsters spend 47 more minutes a day listening to music and other audio.[9] We drive around in cars cocooned in our own portable and controllable soundscape. But then we miss out on simple sonic pleasures: not just the twitter of a bird singing out defiantly against the roar of traffic, the laughter of schoolchildren in a playground, or a snippet of overheard gossip from strangers passing in the street, but the wonderful and unique acoustics of the places we wander through each day. City districts can be visually ugly, but even there, a dingy corner covered in graffiti can harbor the most extraordinary zinging sound effects.

For decades, acoustic engineers have been trying to reduce unwanted noise, but many attempts have been defeated by changes in society. A modern car is much quieter than an old clunker, but increased traffic means the average city noise level has remained about the same. As the rush hour extends and motorists seek out quieter roads, peaceful places and times are disappearing. What should we do about this noise? I believe that telling people to stop doing noisy things is futile. It is better to encourage listening and curiosity. Although technology often produces unwanted noise, new devices are also creating bizarre and wonderful sounds every day. Gadgets emit distinctive pings and buzzes that people will cherish and be nostalgic about. The chime of the pinball machine takes me back to hanging out with friends in my youth. My children will likely have fond memories of the iPhone click, long after the device has been supplanted by more sophisticated technologies. With better awareness of the sonic wonders, I hope people will demand better everyday soundscapes.

Since the trip into the sewer, my hunt for sonic wonders has morphed into a full-blown quest. I set up an interactive website (Sonic Wonders.org) to catalogue my discoveries and serve as a forum for people to suggest enticing sounds for further investigation. After a talk at a conference in London, a delegate told me about a large, spherical room called the Mapparium at the Mary Baker Eddy Library in Boston, where even nonventriloquists can throw their voice. This illusion plays with the mental processing that allows us to locate sound sources, processes that evolved to protect us from predators creeping up on us from behind. A conversation during a TEDx event in Salford made me want to find out more about the moths that have evolved decoy tails to fool echolocating bats. Rummaging through old proceedings from academic conferences has revealed a gold mine of acoustic curiosities, neglected phenomena that devoted scientists have studied alongside their day-to-day research.

Friends and colleagues, and even complete strangers, have given me examples of quirky acoustics and fascinating science. My research has uncovered the ways in which sounds have inspired musicians, artists, and writers: how church acoustics had to be adapted when the liturgy moved from Latin to English, how writers portray subtle acoustic effects such as the indoor feeling at Stonehenge, and how sculptors have ingeniously made sonic crystals that reframe ambient noise.

As a scientist, I want to pick apart what is going on. I went on holiday to Iceland many years ago and marveled at bubbling mud pools, but now I wonder, what caused them to gloop? I have seen Internet videos of a giant Richard Serra sculpture in which a hand clap resounds like a rifle shot. What is happening there? Why does throwing a rock onto a frozen reservoir produce the most astonishing high-pitched twangs? Some of these questions have no readily available answers, but in searching for explanations, I hope to gain insight into how our hearing works, both in these special places and in our everyday lives.

What makes a sound so extraordinary as to be considered one of the sonic wonders of the world? In my quest for these aural gems, I will rely in part on my gut reaction as a trained acoustic engineer: What might be surprising or weird enough to make the experts stop and wonder? One example might be the sonic behavior in an old water cistern in Fort Worden, Washington, which one audio engineer described as "the most acoustically disorienting place I have ever visited."[10] Or perhaps it will be something that takes us back in time to the experiences of our ancestors. Were the Mayan pyramids in Mexico deliberately designed to chirp? Was this sound used as part of their ceremonies? Sonic wonders might also be very rare acoustic effects: Only a few sand dunes sing, droning like a propeller aircraft—a phenomenon that astonished both Charles Darwin and Marco Polo.

Travel guides will be useless. Like most of our texts, they privilege the visual, describing beautiful vistas and iconic architecture while ignoring sounds and unusual acoustics. I was delighted to find the whispering gallery in St. Paul's Cathedral featured in my London guidebook, but this is a rare exception. The whispering gallery appeals to me as a physicist because the movement of sound around the dome fools listeners into hearing mocking voices emerging from the walls.

Music will play an important role in the search, not least because it can provoke strong emotions. Listen to one of Mahler's grand symphonies in an auditorium like the Great Hall of the Viennese Music Association (Wiener Musikverein) in Vienna, and you might feel shivers down your spine. Music is a powerful research tool, used by psychologists and neuroscientists to tweak the emotions of humans to reveal the workings of the brain. Research into music has taught us a great deal about listening—why some things sound nasty or nice, and how evolution has shaped our hearing. Often the best scientific understanding of sound and how we perceive it comes from research into music. But music and speech engage us on a different level. Indeed, recognizable patterns in music and speech can *distract* us from acoustics and natural sounds. This book will thus go beyond music and oratory to discover sounds that are overlooked or neglected.

Inevitably, I will have to use words and analogies from the visual world to describe sonic phenomena; we have relied on the visual for too long for our language to have developed otherwise. A newspaper interview with the artist David Hockney once said something about *seeing* that has stuck with me:

> We don't just see with our eyes, [Hockney] argues, we use our minds and emotions as well. That is the difference between the image which the camera makes—a split-second record from a fixed viewpoint—and the experience of actually looking, of passing through a landscape, constantly scanning and switch-

ing our focus. That is the difference between the passive spectator and the active participant he wants us to become. The latter sees not just geometrically, but psychologically as well.[11]

I want to explore what would happen if we transpose these ideas from the visual to the aural. To see what fascinating sounds reveal themselves, and discover what effect they have on us. This book is about the psychology and neuroscience of hearing as observed and explored by a physicist and acoustic engineer. And no other place embodies the combination of these disciplines better than a concert hall. Strangely, we know more about the human response to classical music in an auditorium than about many more common sounds. Why not, then, start with the most important quality of a concert hall: reverberation?

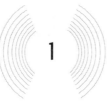

1

The Most Reverberant Place in the World

Guinness recognizes a few world-record sounds: the loudest purr of a domestic cat (67.7 decibels, in case you wondered), the loudest burp by a male (109.9 decibels), the loudest clap measured (113 decibels)—all very impressive. But as a scholar in architectural acoustics, I am more intrigued by the claim that the chapel of the Hamilton Mausoleum in Scotland has the longest echo of any building. According to the 1970 *Guinness Book of Records*, when the solid-bronze doors were slammed shut, it took 15 seconds for the sound to die away to silence.

Guinness describes this phenomenon as the "longest echo," but this is not the right term. Experts in architectural acoustics, like me, use the term *echo* to describe cases in which there is a clearly distinguishable repetition of a sound, as might happen if you yodeled in the mountains. Acousticians use *reverberation* when there is a smooth dying away of sound.

Reverberation is the sound you might hear bouncing around a room after a word or musical note has stopped. Musicians and studio engineers talk about rooms as being *live* or *dead*. A live room is like your bathroom: it reflects your voice back to you and makes you want to sing. A dead room is like a plush hotel room: your voice gets absorbed by the soft furnishings, curtains, and carpet, which dampen the sound. Whether a room sounds echoey or hushed is largely a perception of reverberation. A little bit of reverberation causes a sound to linger—a bloom that subtly reinforces words and notes. In very lively places, such as a cathedral, the reverberation seems to take on a life of its own, lasting long enough to be appreciated in detail. Reverberation enhances music and plays a crucial role in enriching the sound of an orchestra in a grand concert hall. In moderation, it can amplify the voice and make it easier for people to talk to each other across a room. Evidence suggests that the size of a room, sensed through reverberation and other audio cues, affects our emotional response to neutral and nice sounds. We tend to perceive small rooms as being calmer, safer, and more pleasant than large spaces.[1]

I got my chance to explore the record-holding mausoleum at an acoustics conference in Glasgow that included in its program a trip to the chapel. Early on Sunday morning, I joined twenty other acoustic experts outside the gates of the mausoleum. Constructed from interlocking blocks of sandstone, it is a grand, Roman-style building rising 37 meters (40 yards) into the air and flanked by two huge stone lions. An uncharitable observer might try to infer something about the tenth Duke of Hamilton's manhood from the building's shape, which is a stumpy, dome-topped cylinder. It was built in the mid-nineteenth century, but all the remains have long since been removed. The building sank 6 meters (20 feet) because of subsidence caused by mines, which left the crypt vulnerable to flooding from the River Clyde.

The eight-sided chapel is on the first floor and is dimly lit by sunlight shining through the glass cupola. The chapel has four alcoves and a black, brown, and white mosaic marble floor. The original bronze doors that start the world-record echo (modeled on the Ghiberti doors at the Baptistry of St. John in Florence, Italy) are propped up in two of the alcoves. Opposite the new wooden doors is a plinth, built of solid black marble, that once supported an old alabaster sarcophagus of an Egyptian queen within which the embalmed duke was laid to rest. The sarcophagus was actually a bit small for the duke, and our guide delighted in relaying gruesome stories about how the body was shortened to get it to fit. On the day I was there, the plinth was covered in laptops, audio amplifiers, and other paraphernalia for acoustic measurements.

The chapel was meant to be used for religious services, but the acoustic made worship impossible. It was like a large Gothic cathedral, and I found it difficult to talk to my acoustics colleagues unless they were close to me, since the sound bouncing around the chapel made speech muddy and indistinct. But was this the most reverberant place in the world? The record is important to me as an acoustic engineer because the study of reverberation marked the beginning of modern scientific methods being applied to architectural sound.

The scientific discipline of architectural acoustics began in the late nineteenth century with the work of Wallace Clement Sabine, a brilliant physicist who, according to the *Encyclopaedia Britannica*, "never bothered to get his doctorate; his papers were modest in number but exceptional in content."[2] Sabine was a young professor at Harvard University when, in 1895, he was asked to sort out the terrible acoustics of a lecture hall of the Fogg Museum, which (in his own words) "had been found impractical and abandoned as unusable."[3] The hall was a vast, semicircular room with a domed ceiling. Speech in the room was largely unintelligible—a muddy soup of sound more characteristic of the Hamilton Mausoleum than of a well-designed lecture hall. The most forthright critic of the space was Charles Eliot Norton, a senior lecturer in fine arts.

Imagine Norton standing at the front of this vast hall trying to expound on the arts—formally dressed and sporting a large mustache, sideburns, and receding hair. His students would first get the sound traveling directly from the professor to their ears—the sound that goes in a straight line by the shortest route. This direct sound would then be closely followed by reflections—the sound bouncing off the walls, domed ceiling, desks, and other hard surfaces in the room.

These reflections dictate the *architectural acoustics*—that is, how people perceive the sound in a room. Engineers manipulate acoustics by changing the size, shape, and layout of a room. This is why acousticians like me have an uncontrollable desire to clap our hands and hear the pattern of reflections. (My wife was appalled when I clapped my hands in a crypt of a French cathedral. This must go down as one of the more esoteric ways of embarrassing your spouse.) After clapping my hands, I listen for how long it takes the reflections to become inaudible. If sound takes a long time to die away—if it reverberates for too long—then speech will be unintelligible as consecutive words intermingle and become indecipherable. As Henry Matthews wrote in a nineteenth-century text on sound, reverberation "does not politely wait until the speaker is done; but the moment he begins and before he has finished a word, she mocks him as with ten thousand tongues."[4] This is what happened whenever Norton tried to lecture. Students might quip that most lectures are incomprehensible even before the speech is mangled by the room, but Norton was a good communicator and a popular teacher. In this instance it was indeed the fault of the room and not the performer.

Large spaces with hard surfaces, such as cathedrals, the Hamilton Mausoleum, or the cavernous lecture hall at the Fogg Museum, have reflections that persist and are audible for a long time. Soft furnishings absorb sound, reducing reflections and speeding the decay of sound to silence. Wallace Sabine's experiments involved playing around with the amount of soft, absorbing material in the lecture hall—a method that makes him seem like an overenthusiastic fan of scatter cushions. Sabine took 550 one-meter-long (about 1 yard) seat cushions from a nearby theater and gradually brought them into the lecture hall in the Fogg Museum to observe what would happen. He needed quiet, so he worked overnight after the students had gone home and the streetcars had stopped running, timing how long it took sound to die away to nothing. He did not use clapping, maybe because it is difficult to clap consistently unless you are a professional flamenco musician, but instead used a note created by an organ pipe.

Sabine called the time it took for the sound to wither to silence the *reverberation time*, and his work established one of the most important formulations in acoustics. The equation shows how the reverberation time is determined by the size of the room, which is measured by its physical volume, and the amount of acoustic absorbent like the seat cushions from Sabine's experiments or the wall covering of one-inch-thick felt that he ultimately used to treat the acoustics of the lecture hall. One of the crucial decisions engineers make when designing a good-sounding room—whether a grand auditorium, a courtroom, or an open-plan office—is how long the reverberation time should be. Then they can use Sabine's equation to work out how much soft, absorbing stuff is needed.[5]

Alongside reverberation time, a designer has to consider frequency, which directly relates to perceived pitch. When a violinist bows her instrument, the string behaves like a tiny jump rope, whipping around in circles. If she plays the note that musicians call *middle C*, the jump rope turns a full circle 262 times every second. The vibration of the violin radiates 262 sound waves into the air every second, which is a frequency of 262 hertz (often abbreviated to Hz). The unit was named after Heinrich Hertz, the nineteenth-century

German physicist who was the first to broadcast and receive radio waves. The lowest frequency a human can hear is typically around 20 hertz, and for a young adult the highest is about 20,000 hertz. However, the most important frequencies are not at the extremes of hearing. A grand piano has notes from only about 30 to 4,000 hertz. Outside that range we cannot easily discern pitch, and all notes start to sound the same. Beyond 4,000 hertz, melodies are turned into the mindless whistling of someone who is tone-deaf. The middle frequencies where musical notes reside are also where our ears are most efficient at amplifying and hearing sounds. Most speech falls into this range as well, which is why in rooms where music will be played, acoustic engineers concentrate on designing for a frequency range of 100 to 5,000 hertz.

In 2005, Brian Katz and Ewart Wetherill used computer models to explore the effectiveness of Sabine's treatment in the Fogg Museum. They programmed the size and shape of the lecture hall into a computer and employed equations that describe how sound moves around a room and reflects from surfaces and objects. They then added virtual materials to the walls and ceiling of their simulated lecture hall to mimic Sabine's felt treatments. Although the absorbers improved the acoustics, the intelligibility of speech was still poor in places. As one student reported, there were seats where hearing was easy, and conversely, "there were dead spots where hearing was often extremely difficult."[6] Though the treatments were imperfect, Sabine's experiments opened the door for a wide variety of acoustic exploration. His equations remain the foundation of architectural acoustics to this day.

I love walking into a concert hall and hearing the contrast between the small entrance corridor and the huge expansive space of the auditorium. From the claustrophobic passageway, one enters a palpably vast room, passively perceiving the quiet chatter of anticipation among the audience and the occasional loud sound stirring the mighty reverberation. Entering Symphony Hall in Boston is particularly exciting for me. Symphony Hall is Mecca for many acousticians, for it was in this very hall that Wallace Sabine applied his newfound science to create an auditorium still considered to be one of the top three places to hear classical music in the world. Completed in 1900, it has a shoe box shape—long, tall, and narrow—with sixteen replica Greek and Roman statues set into the walls above the balconies. On my visit I settled into one of the creaky black leather seats while the Boston Symphony Orchestra was tuning up on the raised stage in front of the gilded organ. As the first piece began, I could immediately understand why audiences and critics wax lyrical about the place. The hall beautifully embellishes the music, having a reverberation time of about 1.9 seconds.[7] When the orchestra stopped playing at the end of a moderately loud phrase, it took nearly 2 seconds for the sound to become inaudible.

At an outdoor concert, an orchestra might play from a tented stage while the audience enjoys a picnic. The night often ends with bottles of champagne and fireworks exploding overhead. These concerts are fun, but the orchestra sounds thin and remote. In contrast, within a great venue like Symphony Hall the music appears to fill the room and envelop the audience from all sides. The reverberation inside amplifies the orchestra, allowing for more impressive loud playing. It also makes sounds linger a little, enabling musicians to make smoother transitions from note to note. Reverberation helps create a more blended and rich tone. As the twentieth-century conductor Sir Adrian Boult put it, "The ideal concert hall is obviously that into which you make a not very pleasant sound and the audience receives something that is quite beautiful."[8]

The transformational effect of reverberation is not restricted to classical music; it is also used extensively in pop. The 1947 number one hit "Peg o' My Heart" (a slow instrumental played on giant harmonicas), by Jerry Murad's Harmonicats, was the first recording to use reverberation artistically.[9] Since then, "reverb" has become a ubiquitous part of the music producer's tool kit. It makes voices sound richer and more powerful, mimicking what would happen if the person were singing from the stage of a theater. On many television programs where people with lousy voices attempt to sing, as soon as the person hits the first note you can hear the audio engineers slathering on reverberation to rescue the sound.

Reverberation is not the only important feature of a good auditorium. The most infamous concert hall failure is probably the original Philharmonic Hall at Lincoln Center in New York, which opened in 1962 (and was later rebuilt as Avery Fisher Hall). Acoustician Mike Barron describes it as "the most publicized acoustic disaster of the twentieth century."[10] Influential music critic Harold C. Schonberg was particularly vocal, describing the hall as a "great big, yellow, $16,000,000 lemon."[11] Acoustic expert Chris Jaffe described how Schonberg "had a field day writing article after article on the acoustics of the hall as a sort of *All My Children*–type soap opera."[12] Ironically, the acoustic consultant for the hall was Leo Beranek, possibly the most influential architectural acoustician of the twentieth century, and also the only person famous enough to be pursued by groupies at acoustics conferences. I remember my first meeting with Leo over breakfast at a conference when I was a young academic. It was a brilliant chance to talk about my research into concert hall acoustics with this superstar. Unfortunately, he greeted me with a question about why I had been measuring echoes from duck calls (see Chapter 4).

According to Beranek, late alterations to the design doomed Philharmonic Hall. The original concept called for a simple shoe box shape similar to Boston's Symphony Hall. But some thought there were not enough seats in the proposed auditorium. Several New

York newspapers campaigned to have the capacity increased, and Beranek says the committee overseeing the building "caved in."[13] The new design changed the shapes of the balconies and sidewalls, and called for a raft of reflectors above the audience. When the hall opened, critics complained that there was too much treble and not enough bass, and musicians struggled to hear each other, making it difficult for the orchestra to blend its sound. Looking back with current scientific knowledge, Beranek now claims that without these alterations, "we would have been the toasts of New York."[14]

Room shape plays an enormous role in the quality of concert halls. Sound reflections heard from the side are very important because the acoustic waves at our two ears are different. It takes longer for reflections from each side to reach the farthest ear; in addition, that ear is in an acoustic shadow and thus picks up fewer high frequencies because that sound does not easily bend around the head. These two cues signal to the brain that music is not just coming from the stage, but also from room reflections. Because of side reflections, we feel enveloped by the music rather than perceiving the sound as coming from the performers on a distant stage. These reflections also make the orchestra appear physically wider than it is—an effect called *source broadening*, which listeners tend to like.[15] Boston's Symphony Hall achieves this effect through the narrow shoe box shape, which offers plenty of side reflections. The scientific understanding of side reflections has inspired new designs and shapes for halls. Near my home in Manchester, England, the Hallé orchestra plays in the Bridgewater Hall, built in the 1990s. The back half of the audience is divided into blocks with walls between in a pattern called *vineyard terracing*. The partitions separating the audience areas have been carefully angled to create reflections arriving from the side.

Reverberation is all about striking a balance between too little (like being outdoors) and too much. Composer and musician Brian Eno explained the consequences of excess reverberation in the Royal Albert Hall before it was modified:

> It was awful, any piece of music which had rhythm or any kind of speed to it would be completely lost there because every event carried on for long after it was supposed to have ended. It reminds me of when we were at art school, we had a model who was very very fat. We used to say it took her 20 minutes to settle, she was impossible to draw. Well playing fast music in a lot of reverberation is a bit like that.[16]

The desirable amount of reverberation depends on the music that's being listened to. Intricate chamber music by Haydn or Mozart was composed to be heard in courts and palaces, so it works best in smaller spaces with shorter reverberation times— say, 1.5 seconds. The French romantic composer Hector Berlioz wrote about hearing Haydn and Mozart played "in a building far too large and acoustically unsuitable," complaining that they might as well have been played in an open field: "They sounded small, frigid and incoherent."[17]

For romantic music by Berlioz, Tchaikovsky, or Beethoven, more reverberance is needed than for chamber music—say, a reverberation time of about 2 seconds. Organ and choral music demands even more. As renowned American concert organist E. Power Biggs said, "An organist will take all the reverberation time he is given, and then ask for a bit more . . . Many of Bach's organ works are designed . . . to explore reverberation. Consider the pause that follows the ornamented proclamation that opens the famous Toccata in D minor. Obviously this is for the enjoyment of the notes as they remain suspended in the air."[18]

The Royal Festival Hall in London was built as part of the Festival of Britain in 1951, which was meant to help cheer up the nation after years of rationing and austerity during and after the Second World War.[19] While critics adored the building, reviews of the acoustics in the concert hall were mixed, with an eventual consensus forming that the reverberation time was too short, being only 1½ seconds. In 1999, the conductor Sir Simon Rattle said, "The RFH is the worst major concert arena in Europe. The will to live slips away in the first half hour of rehearsal."[20] Hope Bagenal was the original senior acoustic consultant for the hall. Surprisingly, he was not a scientist by training. Acoustic engineer David Trevor-Jones wrote that Bagenal's "broad, liberal education" was important because it gave him "the inquisitiveness and . . . competence to take up as much of the physics of acoustics as he needed."[21] Sabine's equation would have shown Bagenal that there were two solutions to the hall's dry acoustics. The first was to increase the size of the room, allowing more space for the sound to bounce around. Raising the roof would have been effective but impossibly expensive. The second solution was to reduce the acoustic absorption in the room. In a concert hall, the audience is responsible for most of the absorption. Bagenal recommended removing 500 seats to increase the reverberation time, but this was not done.[22] Instead, a revolutionary solution was sought: to use electronics to artificially enhance the acoustics.

In the ceiling of the hall, microphones were mounted inside pots to pick up sound at particular frequencies. The electronic signals from the microphones were then amplified and fed to loudspeakers elsewhere in the ceiling. The sound circulated around a loop, going from microphone to loudspeaker through the electronics, and then from loudspeaker to microphone through the air. This setup sustained the sound for longer in the hall, creating artificial reverberation. This was a remarkable feat of engineering, considering the crude electronics available in the 1960s. The mastermind behind this *assisted resonance* system was Peter Parkin, who began working in acoustics in World War II, helping to defeat underwater acoustic mines. For his work on the Royal Festival Hall, Parkin had a dedicated telephone line from the hall to his house so that he could listen in and check whether the system was working correctly.[23] He was monitoring for faults in the system that might cause the sound circling between the microphones and loudspeakers to grow louder and louder, resulting in feedback, the howling and screeching we associate with heavy metal.

Peter Parkin's electronic system raised the reverberation time from about 1.4 seconds to over 2 seconds for low frequencies, vastly improving the warmth of the sound. But Parkin kept it secret. The use of electronic enhancement with classical music is so controversial that when the assisted-resonance system was first installed, it was brought in gradually without telling the orchestra, audience, or conductors. Once the full system had been covertly used for eight concerts, the engineers then dared to reveal its presence. The system was used until December 1998, when a nonelectronic solution was sought.

I sympathize with the view that classical music should not be electronically enhanced, especially after hearing a demonstration of a different electronic system about twenty years ago in a theater near London. As the engineers switched between different settings, I heard strange mechanical and unnatural distortions, and the sound sometimes even appeared to come from behind me rather than from the stage. Amazingly, this was a demonstration designed to encourage people to *buy* the technology. Nowadays, however, modern digital systems, used in many contemporary theaters, can be impressively effective. I heard one demonstrated last year at an acoustics conference, and with a flick of the switch the lecture hall was transformed into a lyric theater or a grand concert hall with natural-sounding acoustics.

A list of very reverberant spaces would include many mausoleums: the Taj Mahal and Gol Gumbaz in India, Hamilton Mausoleum in Scotland and Tomba Emmanuelle in Oslo, Norway.[24] The large rooms and hard stone walls make these places very lively.

The artist Emanuel Vigeland built Tomba Emmanuelle in 1926 as a museum for his works, but later he decided to use it as his last resting place. Norwegian acoustician and composer Tor Halmrast, who is physically and aurally larger than life, described stooping to enter Tomba Emmanuelle—actually bowing beneath the ashes of the artist, which are in an urn over the door. Halmrast entered a tall, barrel-vaulted room covered with frescoes. He explained, "When entering, you see almost nothing, as the walls are very dark. After some time you see the paintings all over the walls and curved ceiling: everything from life (even copulation before life) until death."[25] One fresco shows a plume of smoke and children rising from a pair of skeletons reclining in the missionary position. The midfrequency reverberation time is 8 seconds, a value you might expect in a very large church, which Halmrast thinks is remarkably long, considering that the room is relatively small.[26]

The explicitly sexual frescoes in Tomba Emmanuelle are in marked contrast to the dour interior of the Hamilton Mausoleum, but which is more reverberant? The world record was based on slamming the bronze doors of the mausoleum's chapel—a very unscientific test. To properly compare the reverberation, one needs initial sounds of equal quality and strength.[27] If Rebecca Offendort from Hilaire Belloc's cautionary tale "Rebecca Who Slammed Doors for Fun and Perished Miserably" were doing a measurement, she would "slam the door like billy-o!" and it would take a long time for the sound to dissipate.[28] A less vigorous experimenter would record a shorter time.

For my visit to the Hamilton Mausoleum, acoustician Bill McTaggart brought along proper measurement gear. Across the

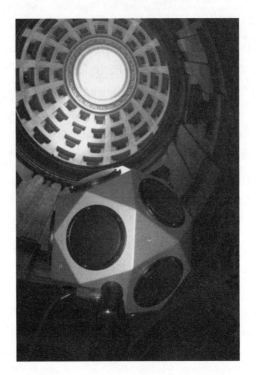

Figure 1.1   Loudspeaker used to test Hamilton Mausoleum, with cupola above.

room he placed on a tripod a strange-looking loudspeaker that sends noise out in all directions (Figure 1.1). It was an icosidodecahedron about the size of a beach ball. A microphone on another tripod stood some yards away. These were all linked to analyzers, whose screens showed graphs of jagged lines sloping from top left to bottom right, characteristic of decaying sound. Normally, acoustic engineers use this equipment to examine whether walls permit too much sound to leak between neighboring houses, or whether the reverberance of a classroom will be too great, undermining the teacher's lessons.

Bill gave a signal, and I quickly stuck my fingers in my ears to protect my hearing. The loudspeakers roared into action, blasting out a growling noise, which even with my ear canals blocked sounded immensely powerful. After 10 seconds, Bill suddenly cut the loudspeaker and measured the noise decay while I quickly unblocked my ears so that I could enjoy the swirl of reverberation. The massive solid walls reflect sound very efficiently, and it took a long time before the noise completely disappeared. The initial overwhelming hissing roar morphed into a rumble that moved above my head, disappearing and dying away up near the cupola. There was a brief moment of silence before the assembled acoustic experts broke into fevered discussions.

What was the reverberation time in the Hamilton Mausoleum? Since it is a large space constructed mainly of stone, the reverberation time is very different at low and high frequencies. At low frequency—say, 125 hertz, which is an octave below middle C (a frequency typical of a bass guitar)—the reverberation time was 18.7 seconds. For midfrequencies it was just over 9 seconds.[29] Impressive, but if it is the longest reverberation in the world, I would be very surprised.

The midfrequencies are where speech is the most powerful, where our ears are the most sensitive, and hence where reverberation times are the most important for clear hearing. No wonder any ideas of holding ceremonies in the chapel were abandoned. Someone talking in a normal way says about three syllables every second. At that rate of speaking in the mausoleum, you utter several words before the first one dies away 9 seconds later. Inevitably, the sounds from many different words mingle and merge. Speech in the chapel is not too bad if you have a conversation with someone close by, because the direct sound from a nearby voice is much louder, making it easier to ignore the reflections. Speaking more slowly also helps. But once you get too far away from someone you want to talk with, the direct sound is less than the reflections, and the reverberation fills in the silences between the syllables, blurring the peaks and troughs in the sound wave and making the speech unintelligible.

Some cathedrals are ten times bigger than the chapel at the Hamilton Mausoleum, and a larger size should mean a longer reverberation time, according to Sabine's equation. The vast, visually imposing cathedrals, built to glorify God, naturally have awe-inspiring acoustics. The sonic qualities are associated with spirituality. The excessive reverberation forces the congregation into silence or hushed whispers, because otherwise speech is rapidly amplified by reflections and creates an ungodly cacophony. During services, the music and words appear to wrap around you like the omnipresent God being worshipped. The acoustic has also influenced services, as the use of chanting and the slow liturgical voice counters the muddy speech in such reverberant spaces.[30]

Many centuries ago, the priest would stand in the chancel, almost cut off from the congregation in the nave. Typically, only a small opening above the chancel screen and below the tympanum would allow sound to reach the churchgoers. The priest would chant facing the altar, his back to the worshippers, so any speech reaching the congregation would be a mush of reflections, with all sound arriving indirectly via reflections off the walls and ceiling. Mind you, with the service mostly in Latin it could be argued it was not the acoustics that caused the speech to be unintelligible.

The Reformation of the sixteenth century changed all that: Anglican priests were instructed by the *Book of Common Prayer* to speak from a place where they could be heard more clearly.[31] Services in English meant that speech would have to be understood. Innovations such as the pulpit in the nave enabled listeners to hear more clearly. There were still reflections, but because the direct speech reached the congregants' ears quickly and some strong reflections arrived shortly after, the setup tended to aid communication. Late reflections, however, make matters worse.

Why are some reflections useful and others detrimental? It comes down to how our hearing has evolved to cope with a complex soundscape. In a cathedral, like most places, the ear is bombarded with reflections from all around—from the floor, walls, ceiling, pews, members of the congregation, and so on. In a large cathedral there are many thousands of reflections per second.[32] Perceiving each individual reflection would quickly overwhelm our hearing. Consequently, the inner ear and brain combine the reflections into a single perceived sound event. Thus, when we clap our hands in a room, we usually hear only one "clap," even though the ear actually receives many thousands of slightly different reflections of the sound in close proximity. A room does not turn a single hand clap into applause.

The ear is a little bit sluggish, rather like a heavyweight boxer. When the ear receives a very short sound, like a hand clap, or when a boxer is hit by a fast punch, it takes a little time for the system to respond to the stimulus. Both the ear and the boxer also continue to respond after the initial stimulus has gone away: the heavyweight boxer reels and rocks for some time after the punch has landed, and similarly, the hair cells in the inner ear continue to send signals up to the brain for some time after the clap has stopped. On top of this physical sluggishness in the ear, the brain is also constantly trying to make sense of the electrical signals coming up the auditory nerves. The brain employs several tactics to separate the priest's direct speech from the morass of late reflections reverberating around the cathedral.[33]

If the priest is to one side, the ear nearer the priest receives louder sound waves because the farther ear receives only speech that has bent around the head. The brain thus attends more to the nearer ear, where the speech is louder and easier to pick out among the reflections. Attention focused in this way becomes less effective if there are lots of reflections from many directions, because both ears become overloaded with a wash of unwanted reverberance.

If the priest is straight ahead, another tactic can be used. In this case the brain adds together what is heard in both ears. The speech coming directly from the priest creates the same signal in both ears because the head is symmetrical, so the sound in each ear has traveled an identical pathway. Adding together the signals from the ears boosts the direct sound. Reflections from the side arrive differently at both ears, and when the left- and right-ear signals are added together, some of the reflections cancel out. This binaural processing increases the loudness of the speech relative to the reverberance.[34]

In big old churches, you often see a small wooden roof (the tester) just above the pulpit. The tester provides beneficial reflections that arrive quickly enough to reinforce the direct sound. The tester also stops the priest's voice from going up to the ceiling to reverberate and return so late that it makes speech less intelligible.

Nowadays, loudspeakers are used to improve speech intelligibility in churches. Like the tester, the loudspeakers direct speech toward the audience, improving the ratio of direct sound to reflections. Older systems used many loudspeakers stacked on top of each other in a line—the idea being that the sound from the loudspeakers adds together to beam the speech toward the audience. More modern systems use sophisticated signal processing to electronically alter the sound coming out of each loudspeaker, creating an especially narrow beam of speech concentrated only on the congregation.[35]

Whereas large churches are something of a nightmare for speech, they make wonderful performance spaces for organ music, as author Peter Smith writes: "The melody line is dominant, but its chords are sounded against the surviving strains of the preceding chords in declining strength. The result is a measure of clash or discord that adds considerable piquancy to the experience. There is a richness . . . in a great cathedral which is absent from the concert hall."[36]

Churches had a profound effect on the development of music. St. Thomas Church (Thomaskirche) in Leipzig, Germany, is an important example. Before the Reformation, the priest's voice took 8 seconds to die away in the church. In the mid-sixteenth century the church was remodeled to help the congregation comprehend the sermons. Wooden galleries and drapes were added that muffled the reverberation, dropping the decay time to 1.6 seconds. Moving forward to the eighteenth century, we find one of the cantors, Johann Sebastian Bach, exploiting the shorter reverberance to write more intricate music with a brisker tempo. Hope Bagenal, the senior acoustic consultant of the Royal Festival Hall in London, considered the insertion of galleries in Lutheran churches, which reduced reverberation, to be "the most important single fact in the history of music because it leads directly to the St Matthew Passion and the B Minor Mass."[37]

How reverberant are grand cathedrals? St. Paul's Cathedral in London was built between 1675 and 1710 to replace the predecessor, which had been destroyed in the Great Fire of London. Designed by Sir Christopher Wren, it has a vast volume of 152,000 cubic meters (5.4 million cubic feet). At midfrequency, the reverberation time is 9.2 seconds; at low frequency it rises a little, to 10.9 seconds at 125 hertz.[38] These decay times are long, but at low frequency the Hamilton Mausoleum is more reverberant, probably because it has fewer windows (which are quite good at absorbing low frequencies). The values at St. Paul's Cathedral are typical of other large Gothic cathedrals, so the mausoleum appears to beat the sanctuary in terms of reverberation.

What about natural spaces, such as caves? The US military became very interested in the acoustics of caves and tunnels during the hunt for Osama Bin Laden in Afghanistan. The idea was to give troops a better understanding of the layout of subterranean passages before they entered. David Bowen, from the acoustic consultancy Acentech, investigated the feasibility by having soldiers fire a gun four or five times at the mouth of a cave and recording the acoustic result. The branches, constrictions, and caverns would alter the way the sound reverberates. This information would reflect back to microphones at the entrance, allowing the cave's geometry to be inferred.[39]

Cave geometry can produce wonderful reverberations. Smoo Cave on the north coast of Scotland emerges among some of the most spectacular and rugged terrain in Britain, with stony green mountains and glorious sandy beaches being bombarded by crashing waves. Nine months after hearing the Hamilton Mausoleum, I went to visit the cave in the hope of finding a more reverberant place. I entered through a large, gaping arch in a sheer limestone cliff cut by the sea. But the first chamber was not as reverberant as I had hoped, because the entrance was very open and there was a large hole in the roof, so the sound quickly disappeared. The second chamber was much more interesting, with a waterfall crashing through a hole in the ceiling, falling 25 meters (about 80 feet) to the flooded cavern floor. The sound was loud and overpowering; when I closed my eyes, it was hard to work out where the noise was coming from as the roar of the cascade reverberated around the cavern.

Basalt columns impressively bedeck Fingal's sea cave on the island of Staffa, Scotland, about 270 kilometers (170 miles) southwest of Smoo Cave. In 1829, the composer Felix Mendelssohn took inspiration from the sound of the Atlantic swell rising, falling, and echoing around the cave. Enclosing the first twenty-one bars of his overture *The Hebrides*, he wrote to his sister Fanny: "In order to make you understand how extraordinarily The Hebrides affected me, I send you the following, which came into my mind there."[40] David Sharp, from the Open University in the UK, measured a reverberation time of 4 seconds in the cave—somewhere between a concert hall and cathedral in the pecking order of reverberance.[41]

In general, although caves can be very large, it seems that the biggest do not reverberate more than grand cathedrals. Writing about a performance of postmodern compositions by Karlheinz Stockhausen in the Jeita Grotto in Lebanon, acoustician Barry Blesser notes that although caves are large, correlating to long reverberation times, they are usually made up of multiple connecting spaces, meaning that the sound decay is "softened, reaching only a modest intensity."[42] Every time a sound wave bounces or reflects, it loses some energy. In a cave, there are lots of side passageways where walls are rough and uneven. The lumps and bumps disrupt the sound, forcing it to bounce back and forth across these passageways and die away faster. The most reverberant spaces have not only smooth walls, but also very simple shapes; this means they are man-made.

In 2006, the Japanese musician, instrument builder, and shaman Akio Suzuki and saxophonist, improviser, and composer John Butcher went on a musical tour of Scotland called *Resonant Spaces*. According to the publicity material, the tour aimed to "set free the sound" of exciting and incredible locations, including the old reservoir in Wormit: "My God it's got a preposterous sound, a huge booming decay and . . . echoes, careening around off its concrete walls. Normally I suppose that would be the worst thing you could think of in a performance venue, but for this tour it's pretty ideal."[43]

An earlier conversation with Mike Caviezel, head of audio for Microsoft games, had piqued my interest in such spaces. After I gave a keynote address at a conference in London, Mike had approached me to tell me about his visit to a similar water reservoir in the US. He described how the acoustics and darkness make it "one of the most crazy, sort of physically disorienting spaces I've ever been in." Mike also described how the reflections affect speaking: "You immediately lose track of what you're talking about, and all you can focus on is just the acoustics of the space." The reverberation is so powerful that "it's very hard to get out . . . clear thoughts or sentences," he said, "and everything quickly devolves to people either whistling, or clapping their hands, or testing the space."[44]

Curious to experience such an odd-sounding space, I decided to visit Wormit a couple of days after I had been in the Hamilton Mausoleum. The arts company that had organized the *Resonant Spaces* musical tour, Arika, gave me contact details for the owner, James Pask, who was only too delighted to show me around. In a gentle Scottish accent, he explained that he had acquired two underground reservoirs when he bought the land; the smaller one had been turned into a vast garage under his house, but the larger one just lay empty underneath his lawn.

We wandered out into the garden, chatting about structural loads and the history of Wormit's municipal infrastructure. The reservoir had been built in 1923 with the intention of serving a large town, but the war intervened and Wormit never grew very large. Eventually, the cost of maintaining the oversized reservoir led it to be decommissioned.

It was very windy that day, with the autumn sun glinting off the Firth of Tay down the hill, and the city of Dundee in the distance across the water. The lawn was extraordinarily flat. Black ventilation pipes poked out of the ground and hinted at what lay below. James opened up a very overgrown manhole cover and asked me if I was worried about health and safety, before disappearing down a ladder into the dark to turn on the light.

The ladders resembled those on ships. The first led down to a

Figure 1.2 Wormit water reservoir (using a very long exposure on the camera).

small platform, and then I had to climb precariously over a chain fence to reach a second ladder, which led to the floor below. The vast space, illuminated by the light streaming through the manhole cover and a single lightbulb, had few visual charms. It was just a concrete box, about 60 meters (200 feet) long, 30 meters (100 feet) wide, and 5 meters (15 feet) high.[45] The concrete on the walls had the texture of the wood shuttering used during the construction imprinted on them (like the walls of the National Theatre in London). It reminded me of a municipal garage, with a forest of concrete pillars regularly spaced about 7 meters (23 feet) apart holding up the concrete ceiling (Figure 1.2). The floor was wet here and there, and it was pleasantly cool, like a natural cavern.

As James and I chatted, the acoustic immediately revealed itself: a rumble began building up and hung about us like a pervasive fog. Many very reverberant rooms are acoustically oppressive, making it hard to have a conversation. But not this reservoir.[46] Surprisingly, we could talk to each other even when we were quite far apart—something that was not possible in the similarly reverberant Hamilton Mausoleum.[47] It reminded me of a cathedral, with the great advantage that I could shout and clap. Whooping unleashed the full power of the "preposterous" acoustic; the sound rattled around for ages before dying away.

I had a few balloons with me, which I burst to get a rough measurement of the reverberation time. As in mausoleums, the most impressive values were at low frequency: 23.7 seconds at 125 hertz. For the midfrequencies that are most important for speech, the reverberation time was a more modest 10.5 seconds.

Saxophonist John Butcher made recordings in the Wormit reservoir as part of the *Resonant Spaces* tour. The *Wire* review of the album describes how he "attacks the spaces."[48] In Butcher's piece "Calls from a Rusty Cage," it is often hard to discern the sound of a saxophone among the strange electronic whistles, breathy squeaks, and blasts, which sound like ship horns. Will Montgomery in *Wire* described how halfway through, Butcher "suddenly leaps into whirling circular breathing with a flamboyant glissando (which . . . recalls the opening to *Rhapsody in Blue*)."[49] This is certainly one musical approach to such a reverberant place: accept the dissonant smog created by lingering notes, and play on.

Another approach is that taken by American trombonist and didgeridoo player Stuart Dempster in his album *Underground Overlays from the Cistern Chapel*. The chapel in question is the Dan Harpole Cistern in Fort Worden State Park, Washington State, the place Mike Caviezel described as crazy and disorienting. It looks very similar to Wormit, although it is circular rather than square. It was built to supply about 7.5 million liters (2 million US gallons) of emergency water for extinguishing fires. A few websites and books quote a 45-second sound decay. This means it takes about 3 seconds for a note to become half as loud, and musicians can achieve note separation only when they play incredibly slowly.[50] *Billboard* magazine described the recording by Stuart Dempster and his fellow musicians as creating an "intensely serene music in which the slightest changes seem cataclysmic, and gradual swells emerge as tidal waves."[51] Writing in the *Times*, Debra Craine describes the music as having an "eerie, majestic calm that envelops you with hypnotic elation."[52] Notes played seconds apart form lush layers on top of each other, requiring a player to think about the interaction of notes played far apart; otherwise, intense dissonance is produced. Stuart Dempster commented, "Usually when you stop for a mistake, the mistake has the decency to stop too, but it doesn't [in the cistern;] it just sits there and laughs at you . . . You have to be a clever composer [or improviser] and incorporate all your errors into the piece."[53]

I listened to the album, enjoying the meditative polyphony, but also listening for the ends of phrases, because after the musicians stopped playing, the sound would naturally ring around the cistern. From these parts of the music, the reverberation time can be estimated. For over a decade, colleagues and I have been developing ways of extracting reverberation time from speech and music. The idea is to make measurements in concert halls, railway stations, and hospitals while they are in use. Conventional reverberation measurements require loud sounds: gunshots or loudspeakers blaring out noise or slow glissandos. These are unpleasant to listen to and can damage hearing. Audience members also have an annoying habit of ruining the results by commenting on the noise— "Wow, that was loud"—as the decay is being measured. But the sound of an orchestra in a concert hall, or the speech of a teacher in a classroom—while imperfect for measurement—does include the room acoustic; the difficult part is finding a way to extract the effects of the room from the music or speech. One of the most exciting areas of research at the moment is the use of computer algorithms to extract information from audio. A well-known example is Shazam, an app that identifies music from a brief recording through a mobile phone's microphone. Other algorithms try to transcribe music automatically or identify the genre of unlabeled audio files.

Applying our algorithm to Stuart Dempster's recording gave an estimated reverberation time of 27 seconds over the low frequency ranges of the trombone and didgeridoo.[54] This is a good indication that the American cistern beats the Scottish reservoir. But to be sure, I wanted a conventional impulse response. When creating a new auditorium, acoustic engineers work from graphs and tables of reverberation times and other parameters to check that the hall meets design specifications. However, these scientific charts and parameters mean little to architects, so acousticians are increasingly making audio facsimiles of a proposed auditorium and getting clients to listen. This *auralization* starts with a piece of music that has been recorded in a completely dead space, like an anechoic chamber (described in Chapter 7). In other words, it is the sound of the orchestra without any room. Acousticians then combine this music with a model of how sound will move in the future place. In the past, impulse responses came from scale models of the auditorium at one-tenth or one-fiftieth of the full size, but nowadays they are more often predicted by computers.

Auralization also works with impulse responses measured in real rooms, so it has also been encoded into artificial reverberation algorithms used by musicians and sound designers creating film and game soundtracks. In one of these reverberators I stumbled across a library of impulse responses that included three measured in the American cistern. At low frequency the Dan Harpole Cistern has the same reverberation time as the Wormit reservoir: 23.7 seconds. But at midfrequency the America cistern wins, with a reverberation time of 13.3 seconds. These reverberation times are longer than those found in even the biggest cathedrals in the world.

Entering the oil storage complex at Inchindown, near Invergordon, Scotland, felt like walking into a villain's secret lair in a James Bond movie. The 210-meter-long (230-yard) entrance tunnel was narrow, concrete lined, and not much taller than me. And as I walked up the gradual incline from the entrance in the hillside, the daylight steadily faded behind and my torch vainly attempted to illuminate the way. The concrete lining ended, the tunnel became bare rock, and an alcove on the left revealed the entrance to the number one oil tank. But this was not a door, because the only way to get through the 2.4-meter-thick (8-foot) concrete wall and into the gigantic storage tank is via one of the four oil pipes, each only 46 centimeters (18 inches) in diameter. This was no time to worry about claustrophobia, because at the other end of the pipes was hopefully the most reverberant space in the world.

I was in Inchindown nine months after experiencing Wormit, visiting tanks that had once held heavy, crude shipping oil. These reservoirs supplied the naval anchorage in the Cromarty Firth at the bottom of the hill. The tanks were constructed in great secrecy amid concerns about the strengthening of Germany's armed forces during the 1930s and the threat posed by long-range bombers, which is why the tanks were dug deep into the hillside. The vast complex took three years to complete. The whole depot held 144 million liters (38 million US gallons) of fuel—enough to fill up two and a half million diesel cars.

My guide was Allan Kilpatrick, an archaeological investigator for the Royal Commission on the Ancient and Historical Monuments of Scotland. Allan is incredibly passionate about the oil tanks, having learned about the secret tunnels as a local boy. With us were about eight other people, taking advantage of a rare opportunity to see the place, although some never got into the main storage tanks, because they found the entrance too claustrophobic.

I was about to enter one of the big tanks, designed to hold 25.5

million liters (about 7 million US gallons) of fuel. I lay down on a trolley, a narrow sheet of metal about 1.5 meters (5 feet) long, and was pushed into the pipe like a pizza being put into a deep oven. The entrance holes looked even smaller when I was waiting to be dispatched, and as I entered I could feel the walls of the pipe pushing into my shoulders, compressing and squeezing me. The helpers kept shoving, my hard hat fell off, and then I was in. It was an undignified landing as I stopped at an angle, with my feet on the floor of the storage tank and my torso still half in the pipe. I struggled upright with a helping hand from Allan—dressed like a climber and looking at home in the dark underground world. Soon afterward, my acoustic measurement gear was pushed through, all carefully selected to ensure it would fit the narrow pipework.

I now had a few moments to take in the tank. All I had was a front bike light, which was too weak to illuminate much of the vast, barrel-vaulted cavern. It was difficult to get a sense of scale. My initial guess that it was 9 meters (30 feet) wide was spot-on. But how high was the space? That was difficult to estimate in the dark. Allan later told me the ceiling was 13.5 meters (44 feet) high.

Much of the floor was covered in pools of water and oil residue. Boots and gloves lay festering in the foul brown liquid, discarded by workers who had the horrible job of cleaning up the oil tanks when they were decommissioned. Fortunately, there was a dry causeway down the middle of the tank following a spine of slightly higher flooring.

As I wandered down the center line, I sang a few notes that hung in the space and built upon each other. In the Baptistry of St. John in Pisa, Italy, there is a long tradition of guides harmonizing with themselves in the impressive reverberance. In the nineteenth century, author William Dean Howells wrote, "The man poured out in quick succession his musical wails, and then ceased, and a choir of heavenly echoes burst forth in response . . . They seemed a celestial compassion that stooped and soothed, and rose again in lofty and solemn acclaim, leaving us poor and penitent and humbled."[55] My singing in the oil tank was much less poetic, I'm afraid, and I contented myself with seeing how many notes I could get going simultaneously—the audio equivalent of a magician spinning plates. There was time to sing great long phrases as the sound seemed to go on forever, maybe half a minute before it died away. The reverberance dwarfed the sound of Wormit's water cistern.

I carried on walking and started to realize how long the tank was: more than twice the length of a soccer or football field, at about 240 meters (260 yards). Whooping brought this giant musical instrument to life. Never before had I heard such a rush of echoes and reverberation. I was like a toddler sitting at a piano for the first time, thrashing the ivories to see what sounds would come out. Reluctantly, after a few minutes I stopped playing with the acoustics and started preparing for my measurements. I put the instrumentation on old heating pipes (used to get the oil flowing), which were covered in a sticky black residue. I fumbled around in the glow of the bike light—tripods under my arms, cables wrapped around my neck, and expensive microphones delicately positioned between my teeth—in a desperate effort not to ruin the equipment.

Modern acoustic measurements are often carried out on laptops, which in theory should make the process easier. But my laptop developed a sense of comic timing: up popped a dialog box announcing that Windows was updating itself deep inside the hillside. I had to resort to plan B: to record gunshots onto a digital recorder.

Allan fired a pistol loaded with blanks about a third of the way into the storage tank, and I recorded the response picked up by the microphones about a third of the way from the far end. This is a standard technique used in concert hall acoustics; old black-and-white photographs show a gun being fired on the stage of the Royal Festival Hall in London when they tested the acoustics back in the

1950s. Although there are many modern measurement techniques using clever noises and chirps, firing a gun is still a respectable and effective technique.

But measuring such a reverberant space was not straightforward. If either I or Allan made a noise—for example, saying to the other something like, "OK, ready to measure"—we would have to wait a minute or more for the sound to die away before we could fire the gun. We also had to stand completely still and make no noise while the sound was decaying, because otherwise the measurement would be ruined. Since we were standing about a hundred yards apart in the pitch black, hand signals were out of the question. Allan suggested we signal by shining torches on the ceiling.

With communication sorted, Allan walked away from me into the gloom. I saw a dim light on the ceiling and responded likewise to show I was ready. The gun went off, and I felt a quick rush of adrenaline as I fumbled with the recorder. But the sound was far too loud, and my digital recorder overloaded. A simple adjustment and I was ready for the second shot, but then I realized I needed to tell Allan what was going on. As I trudged along the center line to explain what had happened, I made a mental note to bring walkie-talkies the next time.

The second shot was fired, and I listened through my headphones, waiting to turn the recorder off when the sound had disappeared. The recording time ticked up on the dial; 10 seconds, 20, 30, 40—still I could clearly hear the reverberation; 50, 60—this was getting ridiculous. After a minute and a half it was completely silent, and I turned off the recorder.

For the third gunshot, I took off my headphones to appreciate the sound. The familiar crack of the gun was followed by a wave of explosion that washed past me and bounced off the end wall, before returning and bathing me in reverberance from all directions. If the world ends with an apocalyptic thunder crack, this is what it will sound like, with the rumble lingering and forlornly dying away. I wanted to shout with astonishment, but I had to remain silent so as not to ruin the recording.

The longevity was extraordinary. The 45-centimeter-thick (18-inch) concrete walls mean that there is very little absorption at low frequency when the sound is reflected from the walls. Furthermore, the shipping oil has clogged up the pores in the concrete, creating a smooth surface impervious to air and thus drastically reducing absorption by the walls at high frequencies. The most absorbent substance in the tank was, in fact, the vast volume of air, which caused a quicker decay at higher frequencies. As the sound wave passed from molecule to molecule, tiny amounts of energy were lost. Textbooks show air absorption of tens of decibels per mile at the highest frequency I measured. In most rooms the distance traveled by sound is too small for this to be important. But the oil tanks are about a sixth of a mile long, so at high frequency the absorption by the air was more important than that by the walls.

With six bangs recorded, it was time for a quick analysis. I transferred the measurements to a laptop and ran my program. My initial reaction was disbelief; the reverberation times were just too long. At this point in relating this story, I like to play a game with my fellow acousticians called "guess the reverberation time." They usually pick an acoustically outrageous number, maybe 10 or 20 seconds. Even so, they always guess far too low. At 125 hertz, the reverberation time was 112 seconds, almost 2 minutes. Even at mid-frequency the reverberation time was 30 seconds. The broadband reverberation time, which considers all frequencies simultaneously, was 75 seconds. I called Allan over to give him the good news. We had discovered the world's most reverberant space.

2

Ringing Rocks

Why did we build huge reverberant cathedrals to celebrate the divine? Did our prehistoric ancestors share our appreciation for resonant spaces? These were the questions going through my mind as I stood by the four tall, imposing façade stones outside a Neolithic burial mound, blowing up party balloons and smiling sheepishly at other tourists. When I bought the balloons, I had been tempted to choose the black one with skeletons printed on them. What could be more appropriate for a burial chamber? But reluctantly, I had settled for some big yellow and blue balloons made from thicker latex, because they would go with a bassy bang.

I had abandoned my bulky acoustic equipment for this field trip. Luckily, I could make surprisingly useful measurements with a pin, a balloon, a microphone, and a digital recorder. I crawled between the entrance stones, and a dank, earthy smell filled my nostrils as I entered the cramped tomb. I set up my microphone in one arm of the cross-shaped chamber, ready to record the balloons bursting on the opposite side.

Only in recent years have scientists begun systematically studying the acoustics of prehistoric archaeological sites. And one of
the more controversial publications on the subject had brought me
to this ancient burial mound only 50 kilometers (about 30 miles)
north of Stonehenge.[1] The whole region is stuffed with prehistoric remains, including the largest prehistoric stone circle in the
world at Avebury, which has 180 unshaped standing stones on a
1.3-kilometer (about ¾-mile) circumference, and Silbury Hill, the
largest prehistoric mound in Europe. At nearly 40 meters (130 feet)
high, made from half a million tons of chalk, the man-made hill
has no clear purpose. But I was measuring a smaller monument,
Wayland's Smithy, a 5,410- to 5,600-year-old Neolithic long barrow (Figure 2.1).

Figure 2.1 Entrance to Wayland's Smithy.

To arrive at the long barrow, I had trudged along the muddy Ridgeway, an ancient walking highway in central England, on a cold, clear winter's day. Had I been on horseback, I could have avoided the quagmire under my feet and also tested the famous legend of Wayland's Smithy, which claims that if you leave your horse tethered overnight along with a silver coin on the capstone, the next morning your steed will be reshod.

The barrow is a large, low mound, fringed by a circle of beech trees. Most visitors poke their heads inside, snap a few pictures, and move on—examining the ancient monument through twenty-first-century eyes. But I felt compelled to explore the acoustics. I listened to my footsteps and how the sound changed as I crawled about. I talked out loud to myself to test whether my voice became distorted, and I clapped my hands to seek out echoes. I even plucked up the courage to sing a few notes, using the acoustics of the burial chamber to enhance my otherwise feeble bass. And of course, I burst my party balloons.

Acoustic exploration is vital for understanding how our ancestors might have used these ancient sites. Back in Neolithic times, sound would have been even more important than it is today. In a time before writing, being able to listen to someone talking, remember the message, and pass it on was a vital skill. Acute hearing was crucial for avoiding predators, repelling attacks from rivals, and tracking and hunting animals for food. To overlook sound is to render the story of ancient monuments incomplete. We need to explore beyond the visual dominance of modern life and use our other senses: hearing, smell, and touch.

An obvious starting point for an exploration of ancient sites is the Greek architectural masterpiece, the theater at Epidaurus. A traveler in 1839 wrote,

I could well imagine the high satisfaction with which the Greek, under the shade of the impending mountain, himself all enthusiasm and passion, rapt in the interest of some deep tragedy, would hang upon the strains of Euripides or Sophocles. What deep-drawn exclamations, what shouts of applause had rung through that solitude, what bursts of joy and grief had echoed from those silent benches![2]

This is a vast, almost semicircular terrace of gray stone seats, banked steeply in front of a circular stage. Even today, tour guides delight in demonstrating the "perfect" acoustics, astonishing visitors as a pin dropped on the stage is heard high up on the huge bank of marble seats. "Few acoustical situations are so enveloped in myth as the antique Greek theatre," wrote acoustic scientist Michael Barron. "For some, the Greeks are credited with an understanding of acoustics which still baffles modern science."[3] Unfortunately, no extant documents reveal what the Greeks knew. But we are not completely bereft of written evidence, because Vitruvius, one of Julius Caesar's military engineers, wrote extensively about the design of Greek and Roman theaters between 27 and 23 BC.[4] What is striking about Vitruvius's book is that the overriding concern is for good acoustics, with less interest shown for visual appearance.

Vitruvius provides simple design principles that still apply today. Greek theaters bring the audience close to the stage so that they can hear the sound as loudly and clearly as possible. This is why the audience seating is roughly semicircular. However, for the audience seated to the side of the stage in Epidaurus, the actors' words would still have been rather quiet, because the voice naturally projects forward.[5] The solution to this problem was to give the side seats to foreigners, latecomers, and women—the ancient equivalent of cheap seats.[6]

Ancient theaters were built in very quiet locations, so that unwanted noise wouldn't drown out an actor's voice. The designs exploited sound reflections, including those from the circular stage floor and scenery. All these reflections reinforced the sound of people speaking on the stage. As the Roman scholar Pliny the Elder noted, "Why are choruses less distinct when the orchestra [stage floor] is covered with straw? Is it due to the roughness that the voice, falling on a surface which is not smooth, is less united, so that it is less? . . . Just in the same way light shines more on a smooth surface because it is not interrupted by any obstruction."[7] The straw probably quieted the sound by absorption rather than scattering. Pliny the Elder's comments are relevant to modern homes, which have become much more reverberant, now that wooden flooring is more fashionable than carpet.

The ancient theaters themselves provide compelling archaeological evidence of an empirical trial-and-error development of good acoustic design.[8] But there is no indication of anything like a modern scientific understanding. Writing about Vitruvius, academics Barry Blesser and Linda-Ruth Salter conclude, "Although some of his insights would be confirmed by modern science, others would prove to be nonsense."[9] The more dubious ideas included the suggestion that a few large vases dotted around the theater would enhance an actor's voice.[10] As a translation of Vitruvius's writings states, "The voice, uttered from the stage as from a center, and spreading and striking against the cavities of the different vessels, as it comes in contact with them, will be increased in clearness of sound, and will wake an harmonious note in unison with itself."[11]

If only acoustic-engineering solutions were that cheap and easy. Unfortunately, the vases would have made little difference to the acoustic. Blow over the neck of a large beer bottle, or more fittingly a large Roman wine jug (say, 40 centimeters, or 16 inches, tall), and you might hear a low, resonant hum. This is the resonant frequency of the air enclosed within the jug. Objects have particular frequencies at which they like to vibrate; flick a champagne flute with a finger and a distinct tone is heard at the glass's natural resonant frequency. But set a wine jug on the floor next to you at Epidaurus, and what you hear is unlikely to change. Any energy used to get the air in the jug resonating will be lost within the vessel. When you walk past empty beer bottles at a pub gig, the sound does not change.

Intriguingly, resonant vases can be found in about 200 churches and mosques built between the eleventh and sixteenth centuries in Europe and in western Asia. These range from 20 to 50 centimeters (8–20 inches) in length, with openings between 2 and 15 centimeters (about 1–6 inches) in diameter. Unfortunately, there are no contemporaneous writings explaining their purpose. High up in the vast Süleymaniye Mosque in Istanbul, you can see a ring of sixty-four small, dark circles just below the ornate ceiling of the dome, which are openings for resonators.[12] In St. Andrew's Church, Lyddington, UK, there are eleven jars high up in the chancel—six in the north wall and five in the south.[13] In the church of St. Nikolas in Famagusta, North Cyprus, holes can be seen that connect to hidden pots and pipes. However, scientific studies have shown they would have been useless.[14] The natural resonant frequencies of some vases do not match the frequencies of speaking or singing, and hundreds of vessels would be needed to have a significant effect.

Such myths probably arise and persist because sound is invisible, so the cause of an aural effect is not always obvious. Before the twentieth-century advent of electronic equipment to record and analyze acoustics, it was impossible to calculate a complicated sound field such as a church. The eminent architectural acoustician Leo Beranek documented some of the myths of acoustics.[15] My favorite is the story of the broken wine bottles found under the stages and in the attics, walls, and crawl spaces of some of the great

European concert halls. Were these artifacts evidence of an ancient technique for improving acoustics, as some have claimed? No, just evidence of the drinking habits of construction workers.

Another myth Beranek notes is the assumption that wooden auditoriums are best because the walls vibrate like the body of a violin. But actually it is better to make the surfaces hard so that sound is not needlessly absorbed. Newer halls that are lined with wood, such as the Tokyo Metropolitan Art Space concert hall, actually use thin veneers of wood glued solidly onto concrete or other heavy and thick substrates.

Greek and Roman theaters are remarkable sonic wonders in which thousands of spectators can hear without the aid of modern electronics. They were clearly designed to achieve good acoustics, but were the Greeks the first skilled acoustic craftsmen?

Sound is ephemeral, disappearing as soon as it's made, so it is difficult to know exactly what our ancient ancestors heard. Evidence of prehistoric acoustics is very sketchy. Musical artifacts provide some of the most robust evidence of our ancestors' sonic world.

The oldest known wind instruments are flutes made from bird bones and ivory, found in a cave in Geissenklösterle, Germany, about 36,000 years old, from the Upper Paleolithic era.[16] The best preserved is made from a hollow vulture's wing bone. It is about 20 centimeters (8 inches) long with a V-shaped notch at one end and five fingering holes.

How can archaeologists be confident that the bones were musical instruments? Holes could be made accidentally; unbelievably, swallowing and regurgitation by hyenas can create round holes in bones.[17] But the Geissenklösterle bones have signs of deliberate and careful working, implying that the holes were precisely and purposefully placed. A replica was made and played. Treating the vulture's wing bone like a flute and blowing over the edge at one end produces a note. Pretending the bone is like a small trumpet and blowing a raspberry down the tube is also effective.[18]

Besides flutes, there is evidence of 30,000-year-old percussion and scraping instruments, along with the prehistoric use of ringing rocks and cave acoustics. A xylophone made from stone might seem an implausible musical instrument, more likely to produce a disappointing clunk than a resonating bong, but certain stones can generate notes. Examples are found around the world: from the tall slender rows of musical pillars in the Vittala Temple in Hampi, India, which ring like bells, to the large rock gongs in the Serengeti, Africa, made from boulders and covered in percussive marks, which make metallic clangs.

Nicole Boivin, from the University of Oxford, has studied the rocky outcrops at Kupgal Hill, southern India. These formations contain boulders of dolerite that create loud ringing tones when hit with granite stones. But did ancient people ever play the rocks? The best evidence is the Neolithic rock art alongside the percussion marks, showing that the site was used for many thousands of years.[19] In a cave at Fieux à Miers in the south of France, there is a large, 2-meter-high (about 7-foot) stalagmite that rings like a gong. Fractures from when it was struck have been dated to 20,000 years ago.[20] Dating percussion marks on rock gongs can be difficult, but in this case the new layers of calcite over the damage give an inkling of the age. What's more, this cave was only recently unsealed, and other prehistoric artifacts found inside indicate when it was occupied.

When I was younger I used to go caving, and I was strictly warned to be very careful of delicate stalactites and stalagmites. Earlier, in the mid-twentieth century, attitudes were more relaxed, allowing an act of "vandalism" to produce the most fantastical stone instrument. Luray Caverns in Virginia contains the Great Stalacpipe Organ, which entertains visitors and occasionally accompanies brides marching down the subterranean isle.

Andrew Campbell, the tinsmith from the town of Luray, discovered the cave back in the late nineteenth century. A report by the Smithsonian Institution in 1880 commented, "There is probably no other cave in the world more completely and profusely decorated with stalactite and stalagmite ornamentation."[21] When I visited, a year after my trip to Wayland's Smithy, I was amazed by the number of formations. They seemed to cover every surface. The curators have lit the cave with bright lights, giving visitors the impression that they're walking around a film set.

The organ is toward the end of the tour. In the middle of the cathedral cavern, among a forest of cave formations, sits an item that superficially resembles a regular church organ. But when a key is pressed, instead of compressed air blowing through an organ pipe, a small rubber plunger taps a stalactite, which rings and makes a note. The current instrument uses stalactites covering 1.4 hectares (3.5 acres) of the cavern. "It is the largest natural musical instrument in the world," the tour guide proudly announced in a staccato Virginia twang so rapid that every other sentence was unintelligible.

With each key connected to a different cave formation, the organ can play thirty-seven different notes. A magazine article from 1957 reports, "Visitors stand enthralled as melody and chords play all round them. No twinkly tunes these, but full-throated music rolling through the cavern."[22] Apparently I heard a rendition of "A Mighty Fortress Is Our God," a sixteenth-century hymn by Martin Luther, but I struggled to pick out any semblance of a tune. It was my own fault; I stood very close to the stalactite that plays the musical note B-flat to get a good view of how it worked. But this meant that the volume balance between the different notes was awry. The cave formations playing the notes are strung out over such a large area that many were too distant and quiet. From where I stood, the music appeared to have only five notes, and it was more like a piece of avant-garde experimental music than a hymn.

In the middle of the cavern the balance between notes is better, and the reverberance of the cave adds an ethereal quality to the music. A combination of the natural ring of the stalactites and reverberance in the cavern means that notes start and end vaguely. By standing close to one stalactite, I could examine the quality of one note in detail. It reminded me of a metallic gong or church bell.

The Great Stalacpipe Organ was the brainchild of Leland W. Sprinkle, an electronic engineer whose day job was at the Pentagon. While visiting the cavern, Sprinkle heard a tour guide hit a cave formation with a rubber hammer, and he was inspired to make the instrument.[23] He then spent three years armed with a small hammer and a tuning fork, searching for good cave formations. When he tapped a stalactite, it would ring with the cave formation's natural resonant frequency. So his task was to find stalactites that produced a beautiful ringing tone and also had a natural resonant frequency close to a note in a musical scale. As Sprinkle discovered, the most visually impressive formations often failed to produce a sound that lived up to their appearance. Only two formations were naturally in tune, so others had to be altered. Sprinkle used an angle grinder to shorten these stalactites, thereby raising the natural frequencies of the cave formations, and eventually he produced a scale of notes that were in tune.

Sprinkle certainly did not spend a long time worrying about appearance. The Stalacpipe Organ looks as if a cowboy electrician botched the cave's wiring. The mechanisms are crudely bolted onto neighboring cave formations and walls, and wires hang loosely and without organization around the space.

Leland Sprinkle is not the only person to become obsessed with making the perfect rock instrument. In the nineteenth century, Joseph Richardson took thirteen years to construct a large stone xylophone out of hornfels slate from the English Lake District. According to the *Journal of Civilization*, Richardson was "a plain unassuming man, with no refinements of education, but possessed of musical talent."[24] The vast instrument currently resides in the Keswick Museum and Art Gallery in Cumbria, where visitors are actively encouraged to play it.

The stones of this "rock harmonicon" span two rows over 4 meters (13 feet) long, with steel bars and bells on two upper levels (Figure 2.2). The bass notes are poorly tuned, and the tone varies across the instrument. Some stones ring beautifully, like a wooden xylophone, while others sound like a beer bottle being struck with a stick. A better percussionist might be able to coax a more musical sound than I did. One historical account recalls, "The tones produced are equal in quality, and sometimes superior in mellowness

Figure 2.2 Richardson's rock harmonicon.

and fulness, to those of a fine piano-forte, under the hand of a skil-ful player."[25] One of the key skills of a good percussionist is the ability to make the mallets rebound quickly, so that they do not inhibit the vibration of the instrument. According to the museum's curator, the whole instrument plays sharp; that is, the frequencies of the notes run higher than the standard scale. To tune the instru-ment, Joseph Richardson chipped away at each slate bar, gradually raising the frequency of the note. If he removed too much stone, the slate played sharp and there was nothing that he could easily do to flatten the note.

According to the *Journal of Civilization*, the Richardson rock har-monicon was so large that it needed three of Joseph Richardson's sons to play it, "one playing the melody, the next executing a clever working inner part, and the third the fundamental bass. Its power extends to a compass of five octaves and a half . . . extending, in fact, as high as the warble of the lark, down to the deep bass of a funeral bell."[26]

I managed a plodding rendition of "God Save the Queen"—quite appropriate, since Queen Victoria had requested command perfor-mances at Buckingham Palace by what a handbill advertising a public concert described as the "Original Monstre Rock Band."[27] According to the *Times*, the first performance was "one of the most extraordinary and novel performances of the Metropolis."[28] The Richardson family toured Britain and the continent playing music by Handel, Mozart, Donizetti, and Rossini.[29]

John Ruskin, the great Victorian writer and critic, used to own a lithophone made from just eight rocks, and in 2010 a new instru-ment was constructed for Ruskin's old home in the English Lake District. Star percussionist Evelyn Glennie gave a celebratory perfor-mance on the new lithophone, which has forty-eight keys arranged in a sweeping arc around the player. The instrument contains green slate, blue granite, hornfels, and limestone from various local valleys and mountains. Writing in the *Guardian* newspaper, Martin Wainwright described the different sounds: "The clinker gives a short, martial note; the green slate a pure, clear, soft sound."[30]

The team of geographers and musicians that constructed this new instrument also investigated what makes a rock ring. The size, shape, and material determine the frequency of the sound. But what intrigues me most is why some stones go bong while others merely clunk. When a percussionist strikes a stone that rings, the energy is held in the rock for some seconds, with the stone's vibration being gradually transformed into sound waves in the air that you hear. The rocks that go clunk lose their energy too rapidly within the stone. Good wineglasses ring when gently tapped. But rest a finger on the edge of the glass, and the sound disappears almost immediately. The friction between the glass and the finger dampens the vibrations and prevents the ringing. For rocks, the damping comes from the internal structure of the stone rather than your finger.

In 2010, I interviewed violin manufacturer George Stoppani for a BBC radio program about how to choose the right wood for the best-sounding violin. He went around his dusty workshop tapping pieces of wood to allow me to hear the different sound qualities. Only wood with the right grain density and microscopic structure produces a clear tone, which rings on for a few seconds—evidence that it can be used to make a world-class violin. It is similar with rock.[31] Within the stone, vibrations are being passed from molecule to molecule. If there are any cracks or hairline fractures, then it is more difficult for the vibration to travel within the rock and the stone will ring less well. In the age of steam, wheel tappers working on the railways exploited the same principle, checking mechanical defects invisible to the naked eye by tapping the wheels of the trains with a small hammer. Lack of a satisfying ring indicated cracks, which could lead to a catastrophic failure of the wheel. But there is more to this than just cracks. Hit a piece of sandstone and it will not ring, whereas a piece of slate, like those I played at the Keswick museum, can impersonate a gong. Both stones originated from layers of sediments, but slate has been transformed by hundreds of millions of years of pressure into a denser material with a more ordered molecular structure. Vibrations can pass more easily between the neatly arranged molecules in slate than between the loosely packed grains of sand in sandstone.

My wife likes to have long phone calls while wandering around the house. As she walks between rooms, her voice changes in fascinating ways, both for her family in the house and for people at the other end of the phone line. Her voice is stronger and harsher in the kitchen because of the hard, reflective tiles and flooring, and clearer and warmer in the living room with soft furnishings, which deaden the sound. The microphone in the handset is picking up a mixture of the sound that travels straight from her mouth and the reflections bouncing off the walls, floor, ceiling, and objects in the room. She cannot sneak into the bathroom during a phone call with me because the bright reverberation is a dead giveaway. Size also matters: larger rooms tend to create a livelier, booming sound.

Now imagine you are prehistoric person wandering around a dimly lit cave system. Your voice will alter as you move from cavern to cavern, through narrow entrances and down tortuous tunnels. The sound quality varies because of the changing patterns of reflections from the rocks. In large caverns a booming reverberance might be heard, in extreme cases mimicking the sound of a church. But in smaller caverns and tight squeezes, the key acoustic effect is *coloration*.

An old staff room at my university had an amazing ability to color sound. It was a plain, narrow, rectangular room with chairs lined up on either side; it was like a waiting room at a train station. The first few times I went into the room I noticed a strange distortion as other people spoke. Moving my head back and forth dramatically changed the timbre of my colleagues' voices. With my head in one position, their speech sounded very bassy and powerful, but elsewhere their voices went all tinny, distorted, and horrible. Colleagues probably wondered if I had been drinking, as I gently swayed back and forth listening to our lunchtime conversations, scientific curiosity trumping self-consciousness.

As I moved my head from side to side, voices in the room changed as if someone was rapidly altering the settings on a hi-fi's graphic equalizer. This coloration was caused by a change in the balance of the sound, with some frequencies being boosted while others were suppressed. It might seem odd to talk about the *color* of a sound, but many of the words we use to describe sounds are appropriated from elsewhere: *bright, warm, dead, live.* The link between color and sound goes back many centuries, with Sir Isaac Newton spotting the similarity between the distance his prism spread out light colors and the lengths of strings needed to sound out a musical scale.[32]

Even today, acoustic engineers carry out measurements using "white" and "pink" noise. When paints are mixed together they form a particular color because the various pigments alter the frequency balance of the reflected light. Blue paint reflects light of a higher frequency than red paint. Similarly, acoustic engineers use colors to describe the dominant frequencies in sounds. White noise contains all frequencies in equal quantities and hisses rather like a poorly tuned radio. Pink noise contains more low frequencies, so it rumbles with a more thunderous quality.

Stairwells with two large, flat, parallel walls are a great place to hear coloration. Just clap your hands and you should hear a shrill, high-pitched note. This is a flutter echo, which is caused by sound bouncing back and forth between the walls, passing your ears over and over again at regular intervals. The frequency of the tone depends on how long it takes for the sound to go from your ear to the walls and back again.[33] If the stairwell is narrow, this round-trip is quick, the reflections from the wall arrive quickly, one after another, and a high-pitched note is heard. For wider stairwells, there is a longer delay between the reflections you hear, and a lower frequency results.

The most extreme flutter echo I have experienced was in *Spiegelei*, a temporary work of art at Tatton Park, Cheshire, England, by artist Jem Finer. This was a spherical camera obscura, a metal sphere about 1 meter (3 feet) in diameter on top of what looked like a large garden shed. Stick your head into the middle of the sphere and you could see images of the park projected upside down onto the inside—the visual distortions being inspired by the artist's memories of taking drugs in the park as a teenager. The exhibition catalogue described the sound inside as "distorted and deranged"—fitting for a work that was playing on the absurdities of gravity.[34] It was fascinating to see how many people experimented with the acoustic once they poked their head inside the sphere. Like a stairwell, the sphere provided sound reflections arriving at regimented intervals. As the curved walls of the sphere focused the sound, the reflections were particularly strong and the coloration especially marked.

You are unlikely to find a perfect sphere in a natural cave. Nevertheless, distinct coloration is heard in caverns. Would prehistoric man have exploited the coloration caused by tight squeezes in caves or the longer-lasting reverberation offered by large caverns? It would be extraordinary if our ancestors had overlooked these effects, especially when you consider how poor the lighting was and how unusual such acoustic effects would have been in an era before buildings. Indeed, starting in the 1980s acoustic archaeologists have been building up evidence that rock art is found in places where the sound is especially noteworthy. One of the pioneers of this work is Iegor Reznikoff:

A remarkable discovery in the study of ornate caves is the relationship between painted red dots in narrow galleries, where one has to crawl, and the maxima of resonance of these galleries. While progressing in the dark gallery, crawling and making vocal sounds, suddenly the whole gallery resonates: you put the light of your torch on, and a red dot is there on the wall of the gallery.[35]

Sound also appears to have influenced what our ancestors painted. Acoustic archaeologist Steven Waller tried to put the work on a more robust scientific footing by statistically analyzing what appears in each acoustic zone. In a paper in *Nature* he wrote, "In the deep caves of Font-de-Gaume and Lascaux, the images of horses, bulls, bison and deer are found in regions with high levels of sound reflection, whereas feline art is found in regions of the caves with poor acoustics."[36] It seems that our ancient ancestors were exploiting cave acoustics as they told stories around their drawings, with tales of loud hoofed animals being amplified by reflections, whereas quiet cats called for no sonic reinforcement.

The sheer volume of evidence that prehistoric rock art was influenced by cave acoustics is quite persuasive. But David Lubman, a retired aerospace engineer who has been applying acoustic science to archaeological sites, warns that correlation does not necessarily mean causation.

I met David in a Vietnamese restaurant in Los Angeles to discuss his work in archaeoacoustics. His wife, Brenda, accompanied us and took the wise precaution of bringing her own car to allow an early escape, because once you get David talking about his favorite subject, it is very difficult to stop him.

"High praise for Dauvois [another researcher] and Reznikoff, and their discovery of that correlation," said David. "I think [this] was a turning point for me."[37] He went on to explain that a proper scientific sound source would have been better than using Reznikoff's voice to test the caves, and that the whole methodology is vulnerable to experimenter bias. David's hypothesis is that the painters chose nonporous rocks for their art because they would be easiest to paint. By chance, nonporous rocks also provide the strongest reflections. Sound waves cannot penetrate an impervious surface; the sounds just bounce off it. In contrast, porous rock has microscopic holes, air channels that the sound waves can enter. In acoustics, air is modeled as a viscous fluid, like molasses, except much runnier. And like molasses, air does not like being forced into the small channels. As sound enters these tiny holes in the porous rock, the vibrating air molecules carrying the sound wave lose energy to heat. Consequently, porous stone provides weaker reflections than do nonporous rocks.

When you come out to a place like this, where it is very quiet and you can hear the echo and imagine what the ancient people were thinking, there is something hypnotic about it, and it really reaches into certain areas of the brain and the soul, and you hear these ancient voices.[38]

This is Steven Waller describing the experience of visiting outdoor ancient rock art. Waller believes that many people miss a trick when visiting prehistoric sites. Not only should we test the sound close to the paintings by clapping, yelling, or singing, but we also need to step back to search for acoustic effects. Stand back from examples in Australia, for instance, and the effect is "almost spooky," he says. "Where they've drawn a person, and you yell at it, it's like the person is speaking to you."[39] A similar effect is heard at Indian Hill, near San Diego, where sound repeatedly echoes from a cave entrance "as if the rock is calling out . . . and spirits were speaking back, right from the place where they chose to decorate."[40] To get these effects, the sound reflecting from the wall or cave must be heard separately from the sound going directly from your throat to your ear. And this happens only if you stand back from the surface so that the reflections are delayed. "Unfortunately, most people walk right up to a painting and study it from inches away, talking in hushed voices," says Waller. "They never step back and see, or hear, the forest for the trees."[41]

I've found rock art difficult to explore as a sonic tourist because many sites have restrictive access to preserve the paintings and some sites have been altered. I was hoping to see if there was an echo at L'Abri du Cap Blanc in France, which is a stunning frieze of prehistoric sculptures carved into a rock shelter. But I was dismayed to have my sonic explorations thwarted by the building that had been constructed to protect the frieze from the elements.[42] One thing that can endanger sonic wonders is well-meaning conservation that considers only the visual to be important.

Waller carried out statistical analysis of Horseshoe Canyon in Utah and Hieroglyphic Canyon in Arizona. The latter is in the Superstition Mountains on the edge of Phoenix, and when I was in the US for the Great Stalacpipe Organ, I took the opportunity to visit it. I set off at sunrise to avoid the worst heat of the day (it peaked at 41°C, 106°F, that day) and admired the stately saguaro cacti dotted over the hillside as I hiked up the 2.4-kilometer (1.5-mile) trail to the Native American rock engravings. The petroglyphs are in a canyon, etched into rocks just above where a stream normally flows (it had dried up when I was there in June). Thousand-year-old geometric shapes—lines of sheep and deer drawn by the ancient Hohokam people—intermingle with more recent graffiti etched by vandals.[43]

Not long after I arrived, I was joined by a large, friendly family whose parents had somehow managed to get their children out of bed very early. Unable to make any acoustic measurements, I sat back and listened to the family playing and exploring the space. As the children yelled, a distinct echo reflecting from the U-shaped mountains could be heard. As they ran about the canyon close to the engravings, their footsteps and high-pitched voices were colored by the reflections from the semi-enclosed rocks. But these effects were not just restricted to the area around the engravings; plenty of unadorned spots had a similar acoustic.

The heat was debilitating, even in the shade. I could not help thinking that however interesting the acoustics are in the canyon, the presence of water must have made this place significant to the Hohokam. The only archaeological study into the canyon that I've been able to find says the spring was a natural place for the artwork because that is where sheep would have gathered to drink.[44]

At Horseshoe Canyon in Utah, the Great Gallery contains particularly fine, often ghostly figures, some of which are even life-sized. These were described by Polly Schaafsma as "dark, tapering, immobile anthropomorphic form[s], painted in dark red pigment ... hovering in rows against a sandstone backdrop within arched alcoves and rock-shelter."[45] Along the canyon, the four sites where the echoes are strongest are those where the paintings are found; Waller's statistical analysis shows that the probability that this co-occurrence arose by chance is one in 10,000.[46] Places with no echoes and good rocks to paint on are not decorated.

Ninety percent of drawings in the Horseshoe Canyon include hoofed animals such as bison or buffalo. Waller has suggested that the percussive echoes mimic the sound of the animals moving and stampeding. Slow-motion footage of horses reveals that two of their feet land on the ground almost, but not quite, at the same time, giving a double "clop." Stand a few tens of yards from a large, flat surface and clap with a steady rhythm, and you can mimic this sound. But you could also produce the rhythm without the echo. When a hoofed animal walks or gallops, there is a distinct jaunty rhythm to the hooves hitting the ground—something I remember simulating with two halves of a coconut when I was a child.

Such theories in archaeoacoustics are necessarily speculative. Some mainstream archaeologists initially doubted David Lubman and his ideas about echoes from Mayan pyramids. As he explained to me, "I thought the archaeologists would be jubilant that somebody has discovered something that they had understandably overlooked, but instead they are angry with me."[47]

The pyramid of Kukulkan, a Mayan feathered serpent deity, at Chichén Itzá in Mexico was built between the eleventh and thirteenth centuries. It is the height of a six-story building, with a square base that is about the size of half a soccer field.[48] On every side it has a staircase running up the middle that contains ninety-one steps, and at the top there is a square temple. Visit the site, and guides will delight in clapping and producing a chirping sound. Stand in the right place, about 10 meters (30 feet) from the bottom of one of the staircases, and reflections from the stairs create a squawking echo with a distinctive descending pitch. David Lubman claims that this echo mimics the call of the scared and venerated quetzal bird.

Imagine an ancient Mayan priest presiding over a ceremony and, with great theatricality, summoning the sound of the quetzal bird by clapping his hands. Did this happen? And is there an even greater story to tell, about how the Mayans built their pyramids with specific acoustics in mind? Is this perhaps another example of their legendary technological talents, now lost?

I will return to the physics of the sound effect in Chapter 4, but for now it is important to know that many staircases can be made to chirp. The Mayan pyramids are not particularly unusual. Rupert Till, a musicologist from the University of Huddersfield, demonstrated this fact while waiting for his *X Factor* audition at the Old Trafford soccer stadium, the home of Manchester United.

Till's studies of ancient acoustics made him curious about whether the steps between the terraces in the stadium would behave like a Mayan pyramid. Sure enough, when he clapped his hands, a distinctive chirp could be heard.[49] Now, no sane person would suggest that the stadium steps were deliberately designed to chirp, so why should anyone assume that the echo from the Mayan pyramid is anything but an acoustic accident, or that it was used during ceremonies?

David Lubman, however, says it's "hard to believe it wasn't intentional, hard to believe it wasn't noticed."[50] And he goes further, explaining that the acoustic phenomenon is connected to the light shadows that are cast on particular days. At the equinoxes, a zigzag shape appears down the side of the staircase—a distinctive shadow forming a tail for the serpent statues at the bottom of the stairs. His explanation is that around the spring equinox, the quetzal bird undertakes a spectacular diving display, and it looks like a flying serpent. At the bottom of the staircase is the head of the serpent, at the exact point you need to clap to make a chirp. So the echo helps explain the visual display.

I think there are three possible scenarios. First, the Mayans deliberately built their pyramids to make serpent shadows and chirping staircases. Second, it was not deliberate design, but the Mayans noticed that their pyramids chirped and then incorporated the sound into ceremonies. Third, and least romantic, modern guides noticed the chirp and made up this tall tale to entertain tourists.

Unpacking which of the three scenarios is correct is difficult. It is like examining the orientation of ancient structures to the stars and sun. It is easy to prove that places are astronomically aligned in an interesting way, but proving that such alignment was deliberate is impossible.[51] We can learn from modern examples where documents exist to resolve arguments. There are whispering galleries in the US, Europe, and Asia where you can hear ghostly voices apparently emerging from within the walls (more on this in Chapter 5). With so many places built that have this acoustic effect, it is tempting to assume deliberate sonic design. But most are accidents of design, and none appear to have been exploited for rituals and ceremonies—even the ones within cathedrals.

I find it hard to believe that Mayan pyramids were deliberately designed to chirp, but I am open to the idea that the sound might have been exploited in ceremonies. Whichever account of events you believe, an important thing to try at Chichén Itzá is to test the chirp and wonder whether, a thousand years ago, Mayan priests did the same, summoning the quetzal bird, a messenger from the gods.

———————

The wind, playing upon the edifice, produced a booming tune, like the note of some gigantic one-stringed harp. No other sound came from it . . . overhead something made the black sky blacker, which had the semblance of a vast architrave uniting the pillars horizontally. They carefully entered beneath and between; the surfaces echoed their soft rustle; but they seemed to be still out of doors. The place was roofless . . . "What can it be?"[52]

This dramatic description of Stonehenge comes toward the tragic end of Thomas Hardy's *Tess of the d'Urbervilles*, where Hardy refers to that famous ring of stones as "a very temple of the winds." Disappointingly, the wind-driven drone has fallen silent, probably because many stones were removed and rearranged in the twentieth century. But even without the "booming tune," the sound of stone circles can surprise because, as Hardy observes, they can have an unexpected indoor quality.

Stonehenge is one of the most iconic prehistoric sites in the world, so naturally it has attracted curious acoustic archaeologists. There have been many suggestions for why ancient denizens built Stonehenge. Ignoring the ridiculous speculation that it might be a UFO landing site, most sensible ideas involve rituals.[53] Across cultures, human rituals involve sound, whether for celebration or for mourning, so it is probably safe to assume that speech, music, and other sounds were played within stone circles.

My colleague Bruno Fazenda and musicologist Rupert Till went through a standard acoustician's ritual of bursting balloons one morning at Stonehenge just after sunrise. Bruno told me that the sunlight catching the mist and clouds cutting through the circle of stones stunned him with its beauty. But he was less impressed by the sound. When he stood in the middle and clapped his hands or burst a balloon, he could hear only a faint echo from the partial remains of the sarsen circle (the iconic uprights that are capped with horizontal stones). Unfortunately, modern-day Stonehenge is very different from its ancient counterpart, and not just because of the polluting noise from the nearby road. With many stones removed or rearranged, the modern acoustic is a poor imitation of any past sonic glories.

To go back in time to hear the past acoustic, Bruno and Rupert decided to travel nearly 8,000 kilometers (5,000 miles) because, extraordinarily, there is a full-scale replica of Stonehenge in Maryhill, Washington. Constructed by a wealthy American named Sam Hill to honor his fallen companions in World War I, the monument was completed when the altar stone was dedicated on July 4, 1918. On a trip to England, Hill had been told about possible human sacrifices at Stonehenge and had decided that a copy of the prehistoric monument would be a fitting tribute to the suffering and loss of servicemen from Klickitat County.[54]

Bruno and Rupert made detailed field measurements at the Maryhill monument one hot and dusty summer, annoying dog walkers and tourists with loud drumming and chirps as they tried to capture and understand the acoustic. They rose early in the morning so that they could get going before the winds had a chance to pick up and create too much buffeting noise on the microphones. Fortunately, the monument was carefully constructed as a faithful replication of one of the old layouts of Stonehenge. Still, some differences between the monument and the original Stonehenge remain: the concrete blocks at Maryhill are too perfectly square and have a finish reminiscent of a 1970s textured ceiling, whereas each Stonehenge stone has individual character from the way the stones were shaped. From my knowledge of designing acoustic reflectors for concert halls, however, I would doubt this made a vast difference to the sound within the circle.

"Maryhill is quite awesome, a beautiful architectural place on the banks of the Columbia River. It is also a valuable archaeological model, a window into the past, giving you a sense of what it would have been like standing in the original Stonehenge," Bruno explained to me.[55] He also described how his footfall on the gravel changed as he entered the circle, giving a striking, unexpected sense of being inside a room. Exactly the same feeling described by Hardy in *Tess of the d'Urbervilles*.

At first the results of the acoustic measurements surprised me. Burst a balloon at Maryhill, and the sound will ring and reverberate for over a second, a decay time more in common with a school hall than an outdoor space. Given that there is no roof and there are spaces between the stones, I naturally assumed sound would rapidly disappear up to the heavens, but actually, some sound stays bouncing around horizontally among the stones. However, the acoustic is subtler than that of a booming school hall because the reflections are quieter; you have to listen carefully to notice the difference. Nevertheless, these reflections would have been helpful during rituals. As Bruno explained, "It is a surprisingly good place for speech, because reflections reinforce your voice and you can talk to people even from behind some of the inner stones."[56]

Whereas the insides of the stones at Stonehenge have been carefully and meticulously worked to make a smoother, more concave shape, the outsides are often quite rough. Pioneering acoustic archaeologists Aaron Watson and David Keating have suggested that the purpose of smoothing the stone interiors might have been to focus the sound.[57] However, Bruno heard no distinct echo caused by focusing from the rings of rocks in the Maryhill monument. The outer sarsen stones might have produced a focused echo on their own, but reflections from the inner ring of stones hid whatever echo there might have been. The ear combines reflections that arrive at roughly the same time. At Maryhill, the sounds bouncing off the inner and outer stones arrive too close together to be perceived separately, rendering inaudible any possible echo.[58] Bruno and Rupert had hoped to hear a whispering-gallery effect around the circle of stones, but the gaps between the stones ruin this effect. Neither did they hear bass drones or Hardy's "temple of the winds"—even when a gale was blowing through the stones in the afternoon.

The chance to experience an ancient resonance like the one that eluded Bruno Fazenda at Stonehenge is what drew me to the Neolithic burial mound at Wayland's Smithy. I also had a less noble motivation: a curiosity to sample a chamber that featured in an infamous academic publication. In 1994, Robert Jahn and collaborators carried out what they described as "rudimentary acoustical measurements" inside six ancient structures.[59] What they found was that the chambers contained acoustic resonances.

Just like a Roman wine jug, a beer bottle contains air with a particular resonant frequency, which is why you can make a flutelike hooting sound when you blow across the top of one. More specifically, when you blow across the mouth of a bottle, a small plug of air in the neck begins vibrating back and forth against an air spring created by the rest of the vessel. Blowing across the mouth of another bottle, identical except for an elongated neck, sounds a lower-frequency note. The elongated neck means that a longer plug of air is oscillating, and because this greater quantity of air is heavier, the resonant frequency is lower.

Paul Devereux, one of Jahn's collaborators, claimed in a 2001 book that the ancient structures had specific resonant frequencies that enhanced the human voice in an intentional way.[60] This assertion irked acoustic scientist and mathematician Matthew Wright, who noted that all enclosed spaces, such as bathrooms or burial chambers, have resonances—rarely as dramatic or obvious as that emanating from an empty beer bottle, but still powerful enough to fool you into thinking you are a great singer while showering. Wright composed a conference paper entitled "Is a Neolithic Burial Chamber Different from My Bathroom, Acoustically Speaking?"[61]

I decided to put Wright's research to the test by analyzing the balloon bursts from Wayland's Smithy and measurements I made in my bathroom (Figure 2.3). Both plots are jagged lines, with distinct peaks and troughs. The peaks represent the frequencies where there are resonances. Any singer will find notes at these frequencies sounding richer and fuller than normal. If you were to sing a note a little above 100 hertz in one of these places, you would excite a resonance that would reinforce and enrich the sound. Move up to about 150 hertz (a musical interval of a perfect fifth, the leap between the first two notes of the *Star Wars* theme), and the valleys in the graphs at that frequency show that there is no prominent resonance to reinforce your voice, so it would sound thinner. The 100-hertz resonance is conveniently located toward the bottom of my vocal range, ideally suited for impersonating Barry White singing "Can't Get Enough of Your Love, Babe" (a song slightly more appropriate in a bathroom than in a burial mound).

The peaks in the plots demonstrate how acoustically similar the bathroom and burial mound are. Burial chambers and bathrooms are similar in size, big enough to get into and lay a body down, whether for disposal of the dead or for a soak in the bath. This means they both have resonances that are in a frequency range useful for enhancing singing.[62]

Matthew Wright's paper concluded that it is unlikely that acoustics influenced the design of burial chambers. After my own scientific exploration, I'm afraid I agree. The cross shape of the Wayland's Smithy has no discernible affect compared to a simple box. Either kind of small room would have blessed our ancestors with resonant frequencies that would have added a booming quality to chanting and singing, if that is what they did around the decomposing bodies of their kin.

Because we listen through twenty-first-century ears, which have

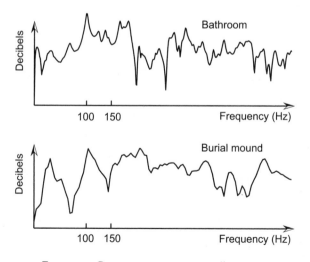

Figure 2.3 Resonances in two small spaces.

become accustomed to hearing reflections from and within buildings most of the time, it is easy to forget how unusual the acoustics of burial chambers and stone circles would have been to our ancient ancestors. Whatever drove the design of Stonehenge, Wayland's Smithy, and other prehistoric sites, we must rediscover our ancestors' listening skills to really understand the archaeology. And that begins with listening to animals.

3

Barking Fish

A year after going to Wayland's Smithy, I joined thirty other people at sunrise of a very cold spring day to hear birds singing the dawn chorus at the Yorkshire Sculpture Park in England. Our guide, Duncan, was a typical bluff and taciturn Yorkshireman who was not going to waste ten words where one would do. "How do you know that's a great tit," I asked. "You just know. Years of listening and watching," was the straight answer. We stood among the trees and sculptures, with bluebells lighting up the undergrowth, all being brought to life by rays of early-morning sun, and just listened. As Duncan might have said, we had signed up for a dawn chorus walk, and listening to birds was all we were going to do.

First I took in the general soundscape. Since it was spring, the birds were in full voice, with song surrounding us from all sides. Duncan was right not to give us expansive descriptions, because being forced to stand and just listen was revelatory, bringing home to me the sheer complexity of the dawn chorus. I tried estimating how many birds were singing and where the sounds were coming from. I tried to pick out individual calls, like a conductor listening for a particular orchestral instrument. In the distance, there was the honking of noisy geese from the lake at the bottom of the hill; they seemed to be constantly screaming. Higher up the slope was the occasional cooing of wood pigeons. In the direction of a rusting sculpture, rooks were cawing their signature call. And all around, songbirds were tweeting and warbling. I picked out one bird singing a powerful short cascade of notes; Duncan identified it as a robin. Now, robins visit my garden all the time, but I had never realized how rich their song is. Chiffchaffs, nuthatches, and chaffinches—how could I have overlooked the diversity of this natural orchestra and just lumped all these little instrumentalists into one big category labeled *birdsong*?

The scientific literature on noise extends even less consideration to individual creatures. A single category contains not only bird calls but also every other natural sound. And there are only two categories: *natural* and *unnatural*. Common sense holds that natural things are good for our health and to be encouraged, and unnatural sounds are harmful and to be abated. But this is an oversimplification that researchers such as Eleanor Ratcliffe, an environmental psychologist from the University of Surrey, are starting to pick apart. Eleanor is researching people's responses to birdsong. In a survey, she found that although birdsong was the most commonly mentioned natural sound, in about a quarter of the responses she read it was unwelcome. For instance, one person complained about the unpleasant, raucous, cackling cry of magpies, partly because magpies have been unfairly blamed for there being fewer songbirds nowadays.[1]

Eleanor has been running other experiments to see whether more pleasant bird calls may be better than others in helping people de-stress. In one test, the call of a small, olive green forest bird from New Zealand, the silvereye, was rated as most likely to help people relax and recover from mental fatigue. The silvereye sings an arche-typal, pretty songbird warble. In contrast, the ugly screech of a jay was rated as less helpful for stress and mental fatigue.

Animal calls are at the heart of our relationship with the natural world. The sounds of insects, birds, and other animals are parts of our memories—evocative of time, place, and season. For me, the croaky "aah" of a rook immediately conjures up images of a churchyard in an English village at twilight, where the birds are settling down to roost. The rhythmic buzz of crickets brings back fond memories of balmy evenings camping in the south of France. When I hear the horrible scream of foxes in heat, I remember wak-ing up once with a start, convinced that a baby was being murdered outside my bedroom window. Many natural sounds are unpleasant like the fox's cry, but could some of these ugly calls be good for us?

Documentary directors portray the natural world as though vision is the only important sense. Sadly, in natural-history televi-sion programs the wildlife sounds are virtually inaudible; instru-mental mood music and pictures dominate. I asked natural-history sound recordist Chris Watson about this. If you have seen any recent BBC natural-history programs, chances are that Chris recorded some of the wildlife. In his soft, northern English drawl, he explained to me that the music is slapped on to manipulate mood: "[It's] so badly done, it's so omnipresent and intrusive, it's like being injected with steroids."[2] But this playing down of natural sound is artificial. How often have you heard wildlife you did not see, because it was difficult to spot or hidden from view? And how did that sound make you feel?

This may not come as a great shock, but science seems to demon-strate that nature is largely good for us. One well-known study showed that patients after gallbladder surgery were discharged from the hospital sooner if their bed had a view out of a window rather than facing a brick wall.[3] Laboratory studies have shown that exposure to nature aids recovery from mental fatigue. Psychologist Marc Berman and collaborators assessed their subjects' mental abilities, for example, by getting them to remember and recount a sequence of digits in reverse order. The subjects then went for a walk in a park or in downtown Ann Arbor, Michigan. After the break, the subjects were retested, and those who had experienced nature outperformed those who had gone downtown.[4]

Nature can also aid recovery from stress. Roger Ulrich and collaborators examined the responses of 120 undergraduate volunteers as they watched a couple of videos.[5] All of the students watched the same first video, which was designed to create stress, depicting accidents in a woodworking shop and including serious injuries, simulated blood, and mutilation. For the second video, half of the students viewed a depiction of a natural setting, and the other half viewed an urban setting. During this second video, the students were asked to rate their affective state while researchers took physiological measurements to gauge how much they were sweating. The students who watched the nature movie recovered more quickly from the stress induced by the accident video than did those who watched the urban recovery film.

Unfortunately, very few studies in this field have focused on the role of acoustics. One rare exception is the study by Jesper Alvarsson and colleagues. They stressed forty people with tricky mental arithmetic tasks. They then allowed the subjects to recover while listening to recordings such as fountains and tweeting birds, or traffic noise, to see how different sounds changed stress recovery. Results were inconclusive, however. Only one of the physiological measures, the amount of sweating, responded positively to the natural sounds.[6]

There are three competing theories as to why nature might be good for us. The first is evolutionary; it suggests that a preference for natural things evolved to encourage us to seek out fertile natural environments where we could find food. The second is psychological and suggests that nature stops us from being too self-absorbed and having negative thoughts, by giving us a sense of belonging to something "greater than oneself." The third theory says that restorative natural places have "soft fascination," meaning that there are captivating yet calming features to look at, such as clouds, sunsets, and the motion of leaves in a breeze, and that this soft fascination helps achieve cognitive quiet.[7] These theories can help explain our reactions to natural sounds that are aesthetically pleasing. But what about those that aren't?

Watching westerns as a child, I always felt that the rhythmic chirring of crickets was ridiculously loud. How could the cowboys sleep with that racket going on? It seemed unbelievable that such a small insect could make such a loud sound.

I got the opportunity to ask some top sound people from Hollywood about this one sunny afternoon, as we sipped margaritas alongside a clear blue swimming pool in Los Angeles. Myron Nettinga is an Academy Award–winning sound mixer and designer. Gregarious, enthusiastic, and fixed with a permanent smile, he explained that insects really are that loud in the midwestern US. But what he said next is what really grabbed my attention. In choosing a cricket to accompany the cowboys eating beans around the fire, a sound designer does not pick out any old recording, but must find the one that portrays the right mood to accent the film's emotional narrative. Myron explained that for a scene about a lazy, calm night in the country, he might pick a soothing cricket, "but hey man, if a guy is creeping around the back of the house looking to jump some people, then all of a sudden there's a cricket and he's agitated and he's a little nervous, and he's stopping and starting."[8] Myron's choice depends on the rhythm of the cricket chirrs and the abruptness with which each sound starts.

While cricket calls vary between species, the various buzzes and chirps all start with stridulation, in which the insect rubs its body parts together.[9] The snowy tree cricket sounds like a quiet trill phone as it rubs its wings together in a rapid scissoring action, swiping a hardened scraper from one wing against the stridulatory file—a structure that looks like a saw blade when magnified under an electron microscope—on the other wing. It is like a tiny version of a scraper percussion instrument you might have played in elementary school. Every time the cricket's scraper hits one of the teeth of the stridulatory file, a small impulsive sound is made. The pitch of the trill depends on how fast the scraper is pulled past the teeth. Typically a tooth on the stridulatory file is hit by the scraper every half millisecond, creating a frequency of 2,000 hertz, a typical pitch for someone whistling.

In a recording I have of a snowy tree cricket (Figure 3.1), the insect draws the wings across each other eight times before pausing for about a third of a second and then doing it again. The snowy tree cricket is nicknamed the *temperature cricket* because its call speeds up as the insect gets hotter. You can estimate the temperature (in Fahrenheit) by counting the number of chirps that happen in a quarter of a minute and then adding 40.[10]

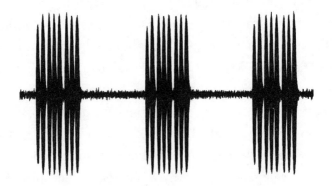

Figure 3.1 The call of the snowy tree cricket.

Given this well-known correlation between temperature and rhythm, sound designers like Myron might choose a slower-chirping, colder cricket to evoke calm (even if the film scene depicts a warmer, lazier night). A hotter cricket sounds more urgent, like a trill phone that demands to be answered, with the chirps closer together and starting more abruptly.

On its own, the stridulation does not make a very loud sound, but each tiny impulsive vibration causes parts of the wing to resonate and amplify the sound. This is similar to how a violin works. The bow vibrates a violin string, which is very quiet on its own. But the string's vibration travels through the instrument's bridge to the wooden body, which has a large surface area and radiates a much louder sound.

The periodic cicada, another insect that uses stridulation, sounds more like a bird than any insect. The cicada's slow, two-tone call starts with an unpleasant, high-pitched screech that lasts a couple of seconds before the frequency drops by about an octave down to a lower, breathy note.[11] The cicada makes short impulses by rapid muscle movements that deform and release tymbal membranes underneath the folded wings—a bit like deforming an aluminum can with your finger. The clicks created when the membrane deforms or pops back are amplified by the resonance of air in the insect's abdominal cavity.[12] While the cicada's dissonant call is terrifying in its own way, I suspect the call is too unusual for sound designers. If you are trying to pull people in and make them feel as though they are really in a film scene, sounds must not be so unusual that they draw attention to themselves. As Myron put it, "You don't want to let them see the magician behind the mirrors . . . You want to make it seem . . . that they're there."[13]

In Bowie, Maryland, near Washington, DC, ash trees filled with a brood of male cicadas can exceed 90 decibels, well above safe workplace levels.[14] Such dense congregation of cicadas happens only every seventeen years, because of the cicadas' long life cycle. The biggest and most common cicada in the Maryland brood was *Magicicada septendecim*, which, according to a report in a local paper, "sounds like a giant [w]eed-whacker or sci-fi spaceship." The loudest cicada in the trees was another species, *Magicicada cassini*, which produces "a harsh screeching noise like the sound of a million baby rattles."[15]

Famous ocean explorer Jacques Cousteau might have celebrated *The Silent World* back in the 1950s, but actually the underwater environment is far from quiet. Water boatmen (*Micronecta scholtzi*) use stridulation to make calls that sound like the rhythmic chirring of crickets. *Micronecta scholtzi* is said to be the loudest aquatic animal relative to its body length. Though only a few millimeters (about a tenth of an inch) long, it can be audible from a riverbank.[16] The discovery that the insect rubs a ridge on its penis against corrugations on its abdomen to initiate the sound made headlines—a rare tabloid triumph for entomological anatomy.

Some water boatmen use the resonance of the air within the bubble they carry around for breathing to amplify their calls. They do this by closely matching the frequency of their body vibrations to the bubble's resonant frequency. As the air bubble shrinks, the resonant frequency rises, and the water boatman needs to stridulate faster.[17]

Snapping shrimp also use bubbles to help make their sound—sometimes for communication, but at other times to kill their prey. The method of making the sound is remarkable because it does not come from the claws tapping each other. In 2000, Michel Versluis, from the University of Twente in the Netherlands, and collaborators used high-speed video to reveal the secret. The shrimp closes its claws very rapidly, with the tips moving at 70 kilometers (about 45 miles) per hour, creating a jet of fast-moving water. Following Bernoulli's principle, the pressure drops in the rapidly moving water, low enough for the water to start boiling at sea temperature. A bubble of water vapor forms, which immediately collapses and creates a shock wave that stuns or kills prey.[18] (Light is also made, in a process, nicknamed "shrimpoluminescence.")

Large colonies of snapping shrimp create a noise like the crackling from a roaring fire. Chris Watson reckons this must be the most common animal sound on our planet, yet "it is a sound that not that many people get to hear."[19] The shrimp also pose problems for natural-history recordists: "I was trying to record the voice and song of the blue whale off the north coast of Iceland, the largest and loudest animal that has ever lived," Chris told me, "and alongside that, at times I couldn't hear the blue whales at a distance because of the snap, crackle, and pop of these animals, which are a couple of centimeters long."[20] This problem is familiar to the military; the study of snapping shrimps began in World War II because the noise was interfering with efforts to hear enemy submarines.[21]

It seems odd that tiny, vulnerable animals draw attention to themselves by making so much noise. The Victorian missionary and explorer David Livingstone wrote on a trip to Africa, "The stridulous piercing notes of the cicadae are perfectly deafening; a drab-colored cricket joins the chorus with a sharp sound, which has as little modulation as the drone of a Scottish bagpipe. I could not conceive how so small a thing could raise such a sound; it seemed to make the ground over it thrill."[22] Perhaps he heard the African cicada? This is the loudest insect, reaching 101 decibels 1 meter (about 3 feet) away—as noisy as a pneumatic drill.[23] But cicadas are not the only incredibly loud chorusing animals. David Livingstone reported, "When cicadae, crickets, and frogs unite, their music may be heard at the distance of a quarter of a mile."[24]

Frogs are meant to go "croak," but someone failed to pass along this information to the amphibians in Hong Kong Park. Built on a former army barracks site in Central District, the park gives respite from one the most densely populated cities in the world. When I visited in 2009, the frogs in the park made squelching chatter like a poor impersonation of Donald Duck. Frogs mostly call with their mouths shut, their vocal sac swelling up beneath the mouth like a giant bubblegum bubble. Frogs do not breathe out when calling for a mate; they circulate the air from lungs to mouth to vocal sac, with the sound escaping via vibrations of their head, vocal sac, and other body parts.[25]

Like humans, frogs have a pair of vocal folds that open and close as the air rushes by, breaking up the constant flow of air into pressure pulses that form sound. Humans amplify their voice using the resonances of the air in their vocal tract (the mouth, nose, and air cavity at the top of the throat). But in frogs, the amplifying resonance comes from the skin of the vocal sac. If a human talks after breathing in helium, the change to a lighter gas in the vocal tract shifts the resonances up in frequency, producing a funny squeaky voice. Get a frog to inhale helium, as some scientists have done, and the call is largely unaltered—evidence that the resonance of the air in the frog's vocal sac is not what amplifies the calls.[26]

Rather than imperiling the community, the communal racket creates an evolutionary defense. While bigger frog choirs attract a few more predators, they also attract a lot more females. Each individual frog is less likely to die, and more likely to find a mate.[27] When I walked too close to the frogs in Hong Kong Park, the croaking suddenly stopped, with the wave of silence signaling a threat through the froggery.

Sound designer Julian Treasure believes that most people find birdsong reassuring because over hundreds of thousands of years we have learned that when the birds are singing, everything is OK. It is when the birds go quiet that you need to worry, because it could be a sign that a predator is about. A plausible argument, but not one, I think, that has ever been tested scientifically.[28] This idea has led Julian to use birdsong in some of his sound designs, including as a crime deterrent in Lancaster, California. Loudspeakers looking like small green bollards dot the flower beds along the main shopping street and are meant to play a mix of twinkly electronic music, water movement sounds, and songbirds.[29] Unfortunately, when I visited Lancaster one Sunday afternoon, the day after I met the Hollywood sound designers, the loudspeakers were just pumping out middle-of-the-road country-and-western music. Not a particularly soothing choice perhaps, but there is precedent for using music as a crime deterrent. In Australia, the "Manilow method" uses easy-listening tracks to disperse teenagers, by making places uncool to hang out in. There is plenty of anecdotal evidence that this tactic works, even though Barry Manilow asked, "Have they thought that the hoodlums might like my music? What if some of them began to sing along with 'Can't Smile without You'?"[30]

There are regular news stories about animal noise disturbances; it is difficult to believe, for example, that people complaining about their neighbor's rooster find this natural sound restorative. While loud animal calls are thrilling to listen to, these sounds overload our auditory system, prevent us from hearing other signs of danger, and can even trigger an early warning system that puts us on alert.

Familiarity plays an important role in our response to all sounds, including animal calls. Andrew Whitehouse, from Aberdeen University in Scotland, has been researching the relationship between birds and people, specifically the effect of birdsong. Early on, the media picked up on his research, prompting people to write to him with their personal stories, which produced a gold mine of data for an anthropologist. Take the following story sent to Andrew, from someone who had moved from the UK to Australia:

> The Australian birdsong is really quite disruptive. We have heard of people emigrating BACK to the UK because of the "ugly" birdsong here. In a nutshell I would describe the sub-

conscious effect of "birdsong" here as being to raise people's tension. It is a series of screeches or other worldly sounds.[31]

There are many stories, like this one, from people who emigrated and were surprised at how much the change in birdsong affected them. Even people who had previously largely ignored natural sounds felt alien because of the bird calls.

Unfamiliarity can also be a source of delight, as I found when I visited the dry rainforests in Queensland, Australia, a few years ago. The eastern whipbird is named after the sound of its call. The male starts with a sustained whistled note for a couple of seconds before an explosive crescendo on a glissando, which abruptly ends like a whip being cracked, leaving the sound to reverberate through the trees.[32] The starting note is at a high frequency, somewhere in the middle of a piccolo's range, and the glissando sweeps across a wide span of almost 8,000 hertz in just 0.17 second, like a piccolo player starting on the lowest note and sweeping through the whole of the instrument's range and beyond.[33] Since the whip cry must be a difficult vocal skill, female whipbirds may well use the quality of the performance as a sign of how fit a male is. Sometimes the song turns into a duet as a female responds with a couple of quick syllables: "chew-chew." The calls are more common when mates are being chosen—a strong indication that the duets play an important part in forming and maintaining partnerships.[34]

Of course, one can hear unexpected bird sounds even in one's home country. Bitterns are reclusive wading birds that, for most of the last century, hovered on the brink of extinction. They are a type of heron that make the most extraordinary bass sounds, which can carry for many miles over their reed bed habitat. Many scientific papers detail how to count and identify individual bitterns from their calls, because they are very difficult to see but easy to hear. Their call is immensely powerful: at 101 decibels at 1 meter (about a yard) away, it has a volume similar to that of a trumpet.[35] And at about 155 hertz, a typical frequency produced by a tuba, the bittern's call is often likened to a distant foghorn.

As sound moves through the air, tiny amounts of energy are lost through absorption every time the air molecules vibrate back and forth, and this absorption limits how far the wave can travel. By definition, lower-frequency sounds vibrate fewer times than high-frequency ones, and thus they lose less energy over long distances and travel much farther than high frequencies. So the bittern's bass booms are effective across the reed beds.

A suffocating thick fog hung over Ham Wall wetland reserve near Glastonbury, England, when I went to hear bitterns one spring morning. We had a very early, five o'clock, start because, like most other birds, bitterns are most vocal at sunrise. My guide was John Drever, who had stuffed his car trunk with strange-shaped microphones, recorders, and boom arms. Wearing a flat cap to protect him from the bitter cold, John looked more like a cat burglar than the friendly musician and acoustic ecologist he really is. Once we had parked at the reserve, we staggered down the path to the nesting sites, barely able to see in the dark fog. We eventually stumbled across a bench in a bird blind, sat down, and listened.

I first heard the call off to the left, sounding like a distant industrial process starting up—quite unlike any other bird I had previously encountered. In *The Hound of the Baskervilles*, the villain, Jack Stapleton, tries to fool Sherlock Holmes by suggesting that a "deep roar" and a "melancholy, throbbing murmur" was a bittern calling and not the hound of hell. Unluckily for Stapleton, a bittern sounds nothing like a dog.[36] The bittern I heard reminded me of someone blowing across a large beer bottle in the pub, or the jug from an old-fashioned jug band. Then a moment later another bittern to the right joined in at a slightly higher frequency. We moved to another blind, where I was close enough to a bittern to hear the buildup to the call. The bird gulped in air four times and then let out seven clear booms a couple of seconds apart. Exactly what the bittern does to make the call is unknown, for this is a secretive and well-camouflaged animal. Rare video footage shows the extraordinary prelude to the booms, as the bittern's throat swells up and the body convulses while the air is gulped in, looking like a cat preparing to cough up a fur ball. But then the bird remains almost still as the sound is made. Since the number of bitterns is growing, maybe more observations of the booming will be possible and will solve the mystery. In 1997 there were only eleven booming males in the UK; in 2012 there were at least a hundred, because of restoration of the reed beds.

Scientists have been measuring when the booms are made to try and understand the purpose of the call and whether it relates to breeding success. The fact that the males call before mating implies that the females assess the fitness of competing males by the strength of their booms. Booms are produced during nesting as well, suggesting that they are also used to defend feeding territories.

An hour and a half after we arrived, the light grew brighter and the bitterns stopped booming. Frozen to the bone, we headed back to the car. I suddenly became aware of the bedlam from the song-birds twittering around me. I had been concentrating so hard on low-frequency booms that I had tuned out the high-frequency war-bling. In this environment, the bittern call could easily be mistaken for a loud human-made noise. For a sound to be restorative, it needs to be unambiguously natural and avoid setting our alert system on edge. If we are familiar with the source or have an expert guide to explain it to us, we can categorize the sound as natural and a non-threat and take comfort in it.

A couple of months before the trip to hear bitterns, at a TEDx conference in Salford, England, biologist Heather Whitney talked about how plants evolved to attract pollinators, such as orchids that look and smell like female wasps to con the males into trying to mate with them and thus spread pollen.[37] It was a great talk, but it was the new acoustic research Heather told me about later in the café that really got me excited. One of her colleagues had found plants that evolved leaves specially shaped to attract their pollinators: echolocating bats.

Beyond the human hearing range is an extraordinary ultrasonic world. Bats exist in a plane of hearing where nearly all sounds exceed 20,000 hertz, or 20 kilohertz (1 kilohertz = 1,000 hertz), the upper threshold of our auditory perception. Three months after the TEDx event, I joined a group of about twenty people for a bat walk at twilight in the moorland village of Greenmount, England. The meeting point was the parking lot of the local pub. It was not difficult to identify the guide. With pictures of flying mammals on her T-shirt and phone, Clare Sefton personified bubbly bat enthusiasm. A research scientist in a different subject, she goes to academic conferences on bats just for fun and is an amateur veterinarian for bats. Before walking down the Kirklees Valley, she showed us a couple of patients she was nursing back to health. One came from the largest species in the UK, a noctule bat, with reddish brown fur, very cute, like a big mouse with wings. It kept opening its mouth and flashing its teeth; as Clare said, it was "having a good look at us," shouting its echolocation signal. The other bat was a tiny common pipistrelle, which, despite having a body only about 4 centimeters (1½ inches) long, manages to eat 3,000 insects in one night.

Since the echolocation calls are at too high a frequency to be heard by humans, we needed some electronic assistance. Clare handed out bat detectors: black boxes about the size of an old brick mobile phone with two controls—one marked *gain*, the other marked *frequency*. As darkness started to fall, our small group of bat hunters set off down a tree-lined path, clutching the hissing detectors. Near an old railway bridge, my detector spurted out a fast series of clicks, like someone rapidly clapping hands in an erratic rhythm. "Pipistrelle," Clare announced, identifying the species from the call pattern. Each click is actually a chirp, a short, sharp yelp descending in frequency. The rate at which chirps are produced changes as a bat approaches an object, to the point where each individual chirp cannot be heard. When this happened, the detector sounded like it was blowing a raspberry.

The next day I examined some recordings of a common pipistrelle. The best way to view each chirp was a spectrogram, because it would show how the frequency of the sound changed over the length of the call. More often used to examine speech, the spectrogram is a wonderful tool for visualizing sounds. In Figure 3.2, the dark descending lines illustrate how the frequency dropped from 70 kilohertz to just under 50 kilohertz over a short (7-millisecond) call.

But how could I have heard this call on the bat monitor, when the frequency is far too high for my hearing? An ultrasonic microphone on the bat monitor picks up the chirps of the bat, and the detector adjusts the tone to be within human hearing range.[38]

Clare was able to identify the bat as a common pipistrelle because each species uses different frequencies to echolocate, producing contrasting sounds from the detector. The noctule bat, for example, produces jazzy, rhythmic lip smacks with a distinct groove. Experts can also tell whether the bat is emerging from its roost, feeding, flying by, or having a chat with a friend from the different sounds of the calls.

I find it particularly remarkable that bats do this with basically the same vocal and hearing apparatus that humans have. To make such high-frequency noise, bats have to push the mammalian body to its extremes. Some bat species produce sound at 200 kilohertz, which means they are opening and closing the gap between their vocal folds 200,000 times a second, although they boast

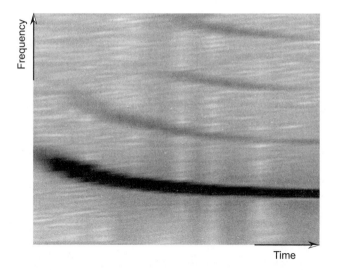

Figure 3.2   The call of the common pipistrelle.

an important modification: thin and light membranes attached to their vocal folds that can vibrate very quickly.

Bats not only hit extremely high notes, but they also routinely generate extraordinarily loud calls. The calls might reach 120 decibels—analogous to the sound reaching your ear from a smoke alarm going off just 10 centimeters (4 inches) away.[39] These are levels that can damage a mammalian hearing system, so bat ears have a reflex to protect themselves: muscles contract when the bat is calling and displace the tiny bones in the middle ear, thereby reducing how much vibration is transmitted from the eardrum to the inner ear. Humans also have this acoustic reflex, but the evolutionary purpose is still being debated. Maybe as in bats, the reflex protects our hearing against loud sounds. Or maybe it reduces the volume of our own speech so that other sounds are more audible.[40]

Leaving the path, we headed through woods toward a small reservoir, stumbling over tree roots as we went (not taking a flashlight along on a nighttime walk was a mistake). But it was worth blundering through the dark to hear the Daubenton's bats hunting insects just above the water. Their roost was under a gigantic brick bridge, and the bat detectors periodically burst to life sounding like distant machine-gun fire. Armed with a detector, I could appreciate the huge number of bats that lived in the valley. It is astounding that up until then I had been totally unaware of these sounds around me. In a radio interview, sound recordist Chris Watson explained how listening to bats hunting down prey changed his perception of Lake Vyrnwy in Wales: "The place was actually turned on its head from being this peaceful tranquil environment to human ears, to being the carnage of a battle above my head in the ultrasonic region."[41]

What else *aren't* we hearing? Marc Holderied at his laboratory in Bristol University, England, is another bat expert with infectious enthusiasm, who answered my questions so extensively that I almost missed my train home. He explained to me that the bats are not only hearing the insects and each other; they are also listening for sound reflecting from plants. Marc and colleagues had been researching *Marcgravia evenia*, a Cuban vine with a leaf that is especially good at reflecting sound and so stands out from the rest of the vegetation in the rainforest. The vine has an arching stalk with a ring of flowers at the end. The last leaf on the branch hangs vertically above the flowers and forms a concave hemisphere to reflect a bat's ultrasonic chirps.

As a bat flies around the rainforest, it hears a very complicated pattern of reflections from all the vegetation. The echoes shimmer and are continually changing. In contrast, the pattern of sound reflections from the convex vine leaf stays almost the same, no matter what the angle of the bat to the plant. So the vine stands out as the only thing in the rainforest that gives a constant reply to the echolocation signal. Moreover, the hemispherical shape of the leaf focuses and amplifies the echolocation signal so that the bat can hear the plant from farther away. Marc and his collaborators confirmed these acoustic properties with laboratory measurements using a tiny loudspeaker to radiate ultrasound and a microphone to sense the reflections off the leaf.

But what evidence suggests that bats take any notice of the reflections from the leaf? By training bats to search for a feeder in a laboratory full of artificial foliage, the researchers demonstrated that the animals found food twice as fast when the hemispherical leaf was in place. In the rainforest, the vine increases its chances of being pollinated by attracting bats with its concave leaf; in exchange, the flying mammals get nectar.[42]

In Marc's laboratory there was a set of dried moth samples, some with extravagantly long tails. Like the vine, these moths have changed because of bats using echolocation. Some moths have evolved high-frequency hearing solely to listen for predator bats. The long moth tails are ultrasonic decoys. A fighter jet can drag a decoy behind it to lure radar-controlled missiles away from the plane. Similarly, a moth sacrifices a decoy tail to protect itself from the bats. The yellow Madagascar moon moth in Marc's laboratory had two swallowtails, each six times the length of the main body. The tails ended in twists, and Marc's measurements show that this means the tails very effectively reflect the bats' calls from all directions, mimicking the ultrasound reflections from the wings of a smaller moth. Marc has shown that 70 percent of the time the bat attacks a streamer tail rather than the insect's body; the moth loses its tail but lives.

Wildlife recordist Chris Watson describes the oceans as "the most sound-rich environment on the planet," adding, "arrogantly we think we're on planet earth, and of course we aren't, we live on planet ocean, 70% of the planet is ocean."[43] To illustrate his point, Chris told me about an expedition to the Arctic, where, off the coast of the island of Spitsbergen in the Svalbard archipelago, he had encountered bearded seals singing underneath thick sea ice. He lowered hydrophones (underwater microphones) through the ice holes made by the seals into the still, inky blackness of the water. To Chris, the seal calls were mesmerizing to listen to because they appeared to be from another planet: "It's almost beyond description. To use lots of clichéd terms, it sounds like a choir of alien angels."[44] The seals make long drawn-out glissandos lasting many tens of seconds. I could do a good impersonation with a Swanee (slide) whistle by gradually pulling out the plunger. It seems that longer glissandos are more appealing to females, so length (of call) matters.

Chris's vivid description of aquatic acoustics made me want to experience these wonders firsthand, which I managed a month after my bat expedition. On a cold, wet, and windy day, I boarded a small boat with a dozen other passengers also swathed in rain gear, clutching a hydrophone and a sound recorder. We were on a trip around the Cromarty Firth in Scotland to meet the resident bottlenose dolphins.

The Cromarty Firth is quite industrialized, and we started by cruising around the gigantic, yellow, rusting legs of an oil rig. In the distance, two other platforms were being repaired, and a cruise liner was moored alongside so that her passengers could hunt for Nessie in nearby Loch Ness. But the dolphins were being as elusive as the fabled monster.

We left the Cromarty Firth and entered the larger North Sea inlet of the Moray Firth. We were close to cliffs flecked with smelly white guano from the seabirds, with green slopes above lit up by swatches of yellow-flowering gorse. Then the skipper, Sarah, caught sight of a dolphin doing an arching jump out of the water.

The engine was switched off to silence its rumble, and I lowered the hydrophone over the side. Initially, all I could hear was the slap of the water on the boat's hull as it bobbed up and down on the swell. Then I heard it—a high-pitched rapid succession of clicks like a tiny toy motorbike revving up, almost inaudible against the water noise.[45]

Next we saw a mother and her calf. The baby dolphin was smaller and a light gray. My thrills were out of sync with the rest of the passengers, as I was the only person with a hydrophone. My fellow passengers were using their eyes to look for dolphins, calling out with excitement whenever one leaped out of the water. But I needed the dolphins to be below the surface for my hydrophones. There was a visual delight having the dolphins so close that I could look them in the eye, but the sound was also magical, because it revealed a little of the underwater world hidden from my fellow passengers.

Unfortunately, human-made noise is forcing animals to change their calls, including underwater mammals and fish. Are offshore wind farms an environmentally friendly way to make electricity? Possibly not, if you are a harbor seal being bombarded with thumping pile driving as the turbines are installed in the seabed. The number of seals counted on the rocks near Great Yarmouth in England declined during the construction of the Scroby Sands offshore wind farm.[46] The noise generated by pile driving is huge—about 250 decibels at 1 meter (3 feet)—and could physically damage the auditory systems of animals.

In March 2000 there was a mass stranding of a dolphin and sixteen whales in the Bahamas that is widely believed to have been caused by US Navy sonar. Scientists dispute how loud sonar causes strandings. The noise may simply cause the whales to swim away, alter their dive patterns, and suffer decompression sickness. Alternatively, the sound waves could cause hemorrhaging. But conclusively proving that naval sonar causes strandings is problematic, because navies are reluctant to say when and where they are using sonar.[47]

A press release published by the environmental lobby group

the Natural Resources Defense Council in October 2005 stated, "Mid-frequency sonar can emit continuous sound well above 235 dB, an intensity roughly comparable to a Saturn V rocket at blast-off."[48] While data show the Saturn V rocket producing 235 decibels, numerically the same as the navy's sonar, the comparison is inapt because of the difference between airborne and aquatic decibels. Similarly, the 250 decibels created underwater by pile driving for wind farms is not the same as 250 decibels in air.

The decibel is always relative to a reference pressure at which you get zero decibels. In air the reference is the threshold of hearing at 1,000 hertz for a healthy young adult. Underwater, the reference pressure is smaller. It is similar to the differences between Celsius and Fahrenheit temperature scales, where 0°C is the freezing point of water, but 0°F is much colder. In addition, when comparing airborne and underwater acoustics, the differences in the density and speed of sound in air and water must be considered. To account for these factors, acousticians subtract 61.5 decibels from underwater measurements to get an equivalent airborne value.[49] Thus, 235 decibels underwater is akin to 173.5 decibels on dry land. In 2008, the *New York Times* described naval sonar as being "as loud as 2000 jet engines," a gross overestimation. The sound 1 meter (about 3 feet) away from a sonar is about as loud as a single jet engine 30 meters (about 100 feet) away—not quiet, but by no means as loud as an entire air force squadron.[50]

Although some of the decibel analogies are dodgy, the gist of the stories about underwater noise causing harm is correct. Many experts are worried because virtually every aquatic animal uses sound as its main way of communicating. Vision is effective only at short distances underwater. Migrating baleen whales can swim more than 100 kilometers (60 miles) in a day, so they need to chat with others in their pod over long distances. Blue whales can be heard from 1,600 kilometers (1,000 miles) away. Whales achieve such long-distance communication by sending out calls at very low frequency, which are much more efficiently transmitted through seawater than are high-frequency vocalizations.

Sudden loud events like naval sonar are not the only sounds that affect marine wildlife. There is chronic shipping noise. In the northeastern Pacific Ocean, shipping noise increased by approximately 19 decibels from 1950 to 2007.[51] This ever-present din could damage aquatic life. It overlaps the frequencies that whales use to communicate, changing vocal patterns: they sing longer, call louder, or move elsewhere. Often whales simply stop communicating, which is a reasonable reaction to short-lived natural sounds, like storms, but not to perpetual shipping noise. Unfortunately, while shipping creates a background cacophony, the noise does not project forward of a ship's bow, which can lead to collisions because whales do not hear boats approaching.

A piece of ingenious scientific opportunism by Rosalind Rolland, of the New England Aquarium in Boston, and colleagues, demonstrated a physiological effect of chronic noise on whales. Rolland's group took advantage of a lull in shipping traffic following the 9/11 terrorist attacks to see how the North Atlantic right-whale population in the Bay of Fundy, Canada, was affected. They monitored whale stress hormones, using sniffer dogs to find floating feces to analyze. After 9/11, the shipping noise reduced by 6 decibels, and Rolland measured a corresponding drop in the whales' stress hormones.[52]

Trying to determine the long-term effects of this chronic noise exposure on sea creatures is difficult. Bombard fish with loud noise in a tank and they will move away. This reaction suggests that noise might displace fish populations from breeding and spawning grounds, and obscure the communication between animals needed for finding mates, navigating, and maintaining social groups. But what scientists are grappling with is how to measure any harm, when the effects might take years to become apparent and aquatic life can move vast distances.

Where does aesthetics fit into whether a natural sound is good for us? In China and Japan, crickets and other insects used to be kept as pets because of their beautiful sounds. In the Sung Dynasty (AD 960–1279), they were the original portable music player. In the introduction of her book on insect musicians, Lisa Ryan writes, "Fashionable people were never without chirping crickets concealed under robes."[53] Rather than pressing a shuffle button, cricket owners used a tickler to stir the insects and make them perform. For me, however, insects are best heard in choruses, especially when the sound can be embellished by the acoustic of a forest. Chris Watson told me about hearing such choruses in the Congo rainforest in Africa. As the temperature drops at sunset, hundreds, possibly thousands, of species contribute to what he described as an "amazing chorus of sound which just rolls out like a wave from the forest."[54] They create a rich musicality, a "Phil Spector Wall of Sound" that within an hour is gone.[55]

Chris's best recordings come from sweet spots where any individual insect is not too prominent and the sound "percolates through the acoustic of the environment."[56] The forest changes the calls, and animals have to adapt in response, to compensate for the distortion created by their surroundings. As sound moves through trees, it bounces off trunks and branches. So, in addition to the direct sound propagating straight from the calling animal, are delayed versions that have reflected off the trees.

The similarity between the acoustic of a forest and that of a room has led to scientific papers with titles such as "Rainforests as Concert Halls for Birds."[57] Recently, walking around lakes and forests in Germany, I tested this out for myself. I noticed how the acoustic changed as I left the open meadows and entered the conifer forest, and when no one was nearby I shouted and listened to the sound reflecting back from the trees. The reverberation time in a forest has been measured at about 1.7 seconds, quite similar to that of a concert hall for baroque music.[58] Forests transmit bass more easily than treble because at high frequency, foliage absorbs sound. This could explain why rainforest birds tend to produce low-frequency songs with drawn-out simple notes.[59] Not only does this form of sound avoid attenuation by the foliage, but the reflections from the trunks amplify the sound of the notes in the same way that room reflections embellish orchestral music. When I shouted in the wood in Germany, I noticed this amplification, but it is subtle because reflections from trees are not as strong as those from the walls of a concert hall.

There is also evidence that birds adapt their singing as their surroundings change. Evolutionary biologist Elizabeth Derryberry examined changes in the calls of male white-crowned sparrows during the last thirty-five years using historical and contemporary recordings from California. In places where foliage has become heavier over the decades, she found that the song pitch has dropped and the birds now sing more slowly.[60] In contrast, the songs remain unaltered in the one area where the foliage has not changed.

Forests are not the only determining factor for birdsong. The most extensive research into chronic noise has investigated how birds deal with the rumble of traffic. Great tits in cities such as London, Paris, and Berlin sing faster and higher in pitch compared with those living in forests; urban nightingales sing louder when there is traffic, and robins now sing more at night, when it is quieter.[61] For great tits, low-frequency singing is important to demonstrate the fitness of males because bigger, healthier birds can sing lower tunes, but their songs can be drowned out by traffic noise. As Hans Slabbekoorn from Leiden University in the Netherlands puts it, "There is a trade off between being heard or being loved."[62] There are fears that noise changes the balance of species, and consequently the songs we hear in cities. It has been suggested that there are fewer house sparrows because they are unable to adapt their songs to the urban din.[63]

The adaptation of song to habitat could be one way that birds develop their own dialects. As humans learn to speak, they pick up an accent as they hear other people talk. Similarly, some bird species learn songs by imitation and so can be influenced by their neighbor's singing. The three-wattled bellbird has different dialects in Central America. One, heard in the northern half of Costa Rica, contains loud bonks and whistles. In contrast, the calls from southern Costa Rica and northern Panama contain loud rasping quacks.[64] Bird dialects have been extensively studied, not least because they give insight into evolution and how species develop. If the songs of neighboring bird colonies diverge—say, because of changing habitats—then eventually they will stop communicating and mating with each other. Once that happens, their genes will no longer mix, which means the colonies will start going down different evolutionary paths and potentially form different species.

The nightingale is a plain-looking bird, but its singing is commonly cited as among the most beautiful in Europe. Listen to a few recordings of a nightingale, and you will notice how many different songs the male can produce. Because they live in thickets, the large vocal repertoire is a more effective display of prowess than is something visual.[65] In 1773, English lawyer, antiquarian, and naturalist Daines Barrington put the nightingale as top-of-his-pops, based on scoring different British birds for their sprightly notes, mellowness of tone, plaintive notes, compass, and execution.[66] A duet between acclaimed cellist Beatrice Harrison and a nightingale was the very first live outside broadcast on BBC radio, in 1924. The nightingales in the woods around Harrison's home in Oxted, England, had taken to echoing her cello practice. The broadcast was almost a failure, however, because the birds were initially microphone shy. But they did eventually sing, and the program was so popular it was repeated for the next twelve years and became internationally renowned.[67]

The nightingale has a beautiful song, which means it is a potentially restorative sound; however, our response to animal calls goes beyond aural aesthetics. When people wrote to Andrew Whitehouse about their experiences of birdsong, the iconic nightingale and its wonderful warbling rarely appeared in the stories. People more often wrote about the stuttering long cry of herring gulls in seaside towns or the excited screeching from flocks of swifts. Sometimes these calls brought back childhood memories: "A Common Gull, just this moment, cried outside my room window. My instant response to that is clear pictures of massed trawlers at Point Law (Pint La) where I spent school holidays." Or songs that signified the seasons: "The birdsong I love the best is the scream of the swift, because of its associations with summer."[68]

Thus, the natural sounds most likely to be restorative and best for our health are those that are familiar and bring back happy memories. When I asked Chris Watson for his favorite sound, he did not choose something exotic from his recording trips around the world; instead he described the complex, rich, and fruity song of the blackbird—something he could hear in his back garden. Hearing nature is different from viewing it, however, so we need new theories to explain which sounds are good for us and why. I enjoy hearing the sound of ducks, not because I think quacks are especially beautiful, but because the sound brings back fond memories of measuring echoes.

4

Echoes of the Past

There is a saying, "A duck's quack doesn't echo and no-one knows the reason why."[1] Hoping to disprove this one slow afternoon at the office, I found myself semiprone on a grassy knoll, pretending to interview a duck named Daisy. Every time she quacked or stretched and opened her wings, camera shutters fluttered like castanets. My colleagues stood close by, unable to contain their laughter. The press had caught wind of our modest attempt to correct the misconception about the supposed non-echoing quack and were doing their best to turn it into an international news event.

Little did I know that, a few years after fronting this frivolous science story, I would once again become engrossed in echoes, rediscovering the childlike pleasure of finding places where a yell resounds with satisfying fidelity to the original. But there is more to echoes than shouting in tunnels or yodeling in the mountains; depending on the type of echo, sound can return magically distorted—claps turning into chirps, whistles, or even zaps from laser guns.

Early documenters of natural phenomena, such as the seventeenth-century English naturalist Robert Plot, used fantastic terms such as *polysyllabical*, *tonical*, *manifold*, and *tautological* to describe the mystery of echoes. But while the cataloguing of animals and birds has survived to the present day and still captures interest, the same is not true of echoes. It is time to revive the echo taxonomy. Can an echo turn a single word into a sentence? Or return the voice "adorned with a peculiar Mu[s]ical note"?[2] Or even transpose a trumpet tune, with each repeat being at a lower frequency?

A few months before the photo shoot with Daisy, Danny McCaul, the laboratory manager at Salford University, had been approached by BBC Radio 2 to find out whether the phrase "a duck's quack doesn't echo" was true or false. Ignoring Danny's careful explanation of why a quack will echo, the factoid was still broadcast. Annoyed that his acoustic prowess had been overlooked, Danny and some of his colleagues, including me, decided we needed to gather scientific evidence to prove the point.

Convincing a farm to lend us a duck and transporting it to the laboratory were probably more time-consuming than the actual experiments. First we placed Daisy in the anechoic chamber and made a baseline measurement of an echo-free quack. The *anechoic chamber* is an ultrasilent room where sound does not reflect from the walls; it is without echoes, as the name implies.[3] It was important to have a reference sound without echoes; after all, this was a serious piece of science and not a bit of Friday afternoon fun. After a brief comfort break for Daisy, she was carried next door to the reverberation chamber, which sounds like a cathedral with a very long reverberation time, despite being little bigger than a tall class-room. Normally, the chamber is used to test the acoustic absorption of building parts like theater seating or studio carpets. In this room, Daisy's quacks sounded evil and ghostly as they echoed around the room, the noise prompting her to cry out again and again. We had created the ultimate sound effect for a horror movie, provided the film featured a vampire duck.

An echo is a delayed repetition of sound, which for a duck might be caused by a quack reflecting off a cliff. The vampiric cry in the reverberation chamber demonstrated that quacks reflect from surfaces like every other sound. We were not surprised by the result, not least because there are bird species that echolocate, using wall reflections to navigate caves. The great Prussian naturalist and explorer Alexander von Humboldt wrote about one of these species, the oilbird, a nocturnal frugivore (fruit eater) from South America. On a visit to the Guácharo Cave in Venezuela in the late eighteenth century, Humboldt experienced the squawking and clicking of the roosting birds. The clicks are the echolocation signals; the birds listen to the reflections to navigate in the dark.[4]

But caves and reverberation chambers are not a natural habit for ducks like Daisy. We were curious to know what happens outdoors. To hear a clear single echo from Daisy, I would need a stretch of water with a large reflecting surface, such as a cliff, nearby. In such a place, sound would travel directly from the duck to my ear, followed shortly by the delayed reflection from the cliff. In the taxonomy of echoes, this is a *monosyllabic echo*, where there is just time to say one syllable before an echo arrives. But Daisy and I could not be too near the cliff, or my brain would combine the reflection with the quack traveling directly from her beak to my ear, and I would hear only one sound.

I must admit that my field experiments were crude. Though I could not bring Daisy, I did wander around various ponds, canals, and rivers listening to wildfowl. In none of these places could I hear a clear, audible quack separate from the original call. In the end, I came to the conclusion that the phase should say, "A duck's quack might echo, but it's impossible to hear unless the bird quacks while flying under a bridge."

Maybe I should have taken Daisy to Lake Königssee in Bavaria, Germany's highest lake, where rock faces rise steeply out of the water. Boat captains there play short phrases on trumpets so that tourists can hear the last three notes repeat, delayed by one or two seconds, after bouncing off the surrounding alps. Or perhaps I should have taken Daisy to the place where seventeenth-century French theologian, natural philosopher, and mathematician Marin Mersenne had carried out his echo experiments. He used a *polysyllabic echo* to make the first accurate measurements of the speed of sound in air. Nowadays, Mersenne is probably best known in mathematics for his work on prime numbers, but he was also passionate about a wide variety of subjects and devoted to the need for experimentation and observation.[5]

Unsurprisingly, Mersenne did not use wildfowl in his speed-of-sound experiments. Instead, he stood facing a large reflecting surface, saying the words "benedicam dominum," and using a pendulum to time the sound. Mersenne must have been a quick talker, because he said this seven-syllable phrase in one second. When he stood 485 royal feet (159 meters) from a large reflecting surface,[6] the echo immediately followed the end of the original phrase—"benedicam dominum, benedicam dominum." This is a polysyllabic echo because many syllables can be said before the echo returns. The echoed words had covered a round-trip of two times 485 royal feet (a total of 319 meters), allowing Mersenne to deduce that 319 meters per second is the speed of sound. This is remarkably close to the correct value of about 340 meters per second (761 miles per hour).[7]

Now, if Mersenne had used a duck, he could have stood nearer the wall and still heard a distinct "quack, quack" because a duck's call is but one syllable. In fact, the distance to hear a monosyllabic echo like a quack is about 33 meters (36 yards) from a reflecting surface (660 duck feet?[8]) because at that distance an echo takes just long enough to bounce back that it can be heard separately from the original sound. To hear a quack echo, I would need to find a stretch of water with a large building or cliff about 30–40 meters (or 30–40 yards) away. But even this would not work, because a duck's quack is too quiet. Sound becomes quieter the farther you are from a source, by 6 decibels for every doubling of distance, so if a quack measures 60 decibels 1 meter (3¼ feet) from the beak, 2 meters (6½ feet) away it will have dropped to 54 decibels, 4 meters (13 feet) away it will be at 48 decibels, and so on. By the time the reflected quack has undertaken its round-trip of 66 meters (72 yards), the echo will be about 24 decibels. In a completely silent place a human could hear this, but more often other noises, such as the distant rumble of traffic or wind moving through trees, are louder, making the quack inaudible.[9] Sadly, even in a silent place Daisy would not be able to hear the echo, because her hearing is less sensitive than a human's. So the reason the echo from a duck's call is not heard is pure physics: a quack is not loud enough to be heard after it makes the return trip from the required distance.

Marin Mersenne's acoustic work extended beyond the speed of sound; he also debunked fanciful tales about 400 years before myth busting became a popular pastime on television. One of the more extravagant acoustic claims in classical literature is a supposed *heterophonic echo* that, when spoken to in French, replied in Spanish. Marsenne knew this could not be true, but as Professor Fredrick Vinton Hunt wrote in his seminal book *Origins in Acoustics*, Marsenne "almost convinced himself that one could devise a special series of sounds whose echo might lead a listener to think he had heard the response in a different language."[10] The term *heterophony* comes from musicology and denotes a melody being simultaneously played with an elaborated variant, so I can only imagine that a heterophonic echo might augment the French words to make them Spanish. Unfortunately, no one knows for sure what was meant by the term, and there are no examples of heterophonic echoes. Luckily, there are other literary games that can be more successfully played with echoes, as I found out in France.

One hot, sunny day in 2011, I was cycling with my family in the Loire valley, and we arrived at the Château de Chinon. The heart of the castle was built by Henri Plantagenêt, who would later become King Henry II of England. But I was more interested in a very unusual road sign just outside the castle walls. It points up a small lane and simply says, "Écho." How could a collector of sonic wonders resist this invitation? A few hundred yards up the lane was a small, raised turnout and a sign indicating that this was the place to test the acoustics. I yelled and yodeled and appreciated the fine echo.[11] What made the experience very satisfying was that the side of the château, which was reflecting the sound, was partly hidden by an orchard, so the clarity of the echo was surprising. I could not resist attempting a traditional piece of echo humor from my guidebook to the Loire:[12]

> Me: "Les femmes de Chinon sont-elles fidèles"
> Echo: "Elles?"
> Me: "Oui, Les femmes de Chinon"
> Echo: "Non!"

This translates into English as:

> Me: "Are the women of Chinon faithful?"
> Echo: "Them?"
> Me: "Yes, the women of Chinon"
> Echo: "No!"

Given the right enunciation, with an unnatural stress placed on the last syllable in each phrase, such as the "non" in *Chinon*, the rhyme worked. By that I mean that the partial words echoed back from

the north side of the château and were clearly audible. The senti-
ments expressed in the poem could not be readily verified.

There are other echo stories. Here is a nineteenth-century account
from *Wonders of Acoustics* by Rodolphe Radau (with a Latin translation in square brackets):

> Cardan tells a story of a man who, wishing to cross a river,
> could not find the ford. In his disappointment he heaved a
> sigh. "Oh!" replied the echo. He thought himself no longer
> alone, and began the following dialogue:
>> Onde devo passar? [Hence I have to pass?]
>> Passa. [pass]
>> Qui ? [here?]
>> Qui. [here]
> However, seeing he had a dangerous whirlpool to pass, he
> asked again
>> Devo passar qui? [I have to pass here?]
>> Passa qui. [pass here]
> The man was frightened, thinking himself the sport of some
> mocking demon, and returned home without daring to cross
> the water.[13]

Wonders of Acoustics includes much about Athanasius Kircher, a
seventeenth-century Jesuit scholar based in Rome who wrote extensively about theater acoustics and other marvels. He was intrigued
by *manifold echoes*—echoes that produce multiple distinct reflections. Included in this category are repeating echoes caused by elaborate structures that turn one word into a whole sentence. For his
two-volume masterpiece *Musurgia Universalis*, from 1650, Kircher
produced a drawing of large upright panels, spaced at various distances from a talker, to generate a series of reflections arriving one
after another. One such edifice had five panels and was designed

to take the word *clamore* and break it down to echo *clamore* from the first panel and *amore* from the second, followed by *more*, *ore*, *re* from the third, fourth, and fifth surfaces, respectively. So, if you shouted the question "*Tibi vero gratias agam, quo clamore?*" ("How shall I cry out my thanks to thee?"), the echoes from the last word would reply with a Latin phrase "*clamore, amore, more, ore, re*," which roughly translates as "with thy love, thy wont, thy words, thy deeds."[14]

I thought this seemed very unlikely to work, but the idea was intriguing enough to inspire a quick test. Not having five large panels lying around, I decided to try out the idea by simulating the situation on my computer. I recorded myself saying "*clamore*" and then, using a piece of prediction software, estimated the reflections returning from each of the panels shown in Kircher's picture. I played around with how far the panels were from the speaker and the loudness of the reflections in an attempt to produce the echo pattern.[15] Much to my surprise, the echo phrase actually worked, but perhaps only because my brain was fooled into hearing patterns I wanted to hear.

I once saw a great demonstration of a similar effect by author Simon Singh, based on the accusation that Led Zeppelin hid satanic messages in "Stairway to Heaven." If you play the track backward, you supposedly hear, "Oh here's to my sweet Satan. The one whose little path would make me sad, whose power is Satan. He'll give those with him 666, there was a little toolshed where he made us suffer, sad Satan." So concerned were some religious groups that various US states introduced legislation requiring records to carry warning labels.[16] These claims further implied that even when records were listened to normally, with the sound playing forward, the listener would subconsciously decipher the meaning of the backward satanic messages.[17]

Several groups of psychologists have tested the claims using proper scientific methods. Experiments showed that if you listen to "Stairway to Heaven" backward with your eyes closed, what you actually hear is gibberish. These satanic lyrics are heard only if you have a printed version in front of you. (You can try this out yourself; plenty of websites are dedicated to *backward masking* with sound samples.) The brain has to make sense of incomplete information all the time, so it is very adept at finding patterns and fitting together different sources of information. But sometimes the brain gets it wrong, in this case matching the written lyrics to the otherwise incomprehensible backward murmurings.

The same thing happens with the "*clamore, amore, more, ore, re*" echo. When I listened very carefully for this pattern of words, I could pick out the phrase. The effect was especially strong if the echoes were faint and I was forced to strain to hear them. But if I closed my eyes and listened more holistically and analytically, the dominant effect I could hear was many repetitions of "*re.*" The clever wordplay disappeared.

A multiple echo, or *tautological echo*, is almost the same as a manifold echo, except the same words or syllables are repeated many times. An episode of the TV show *The Simpsons* features this phenomenon as a piece of aural slapstick. Marge is in church and yet again is being embarrassed by Homer. She cries out, "Homer, your behavior is heinous" and a tautological echo replies "anus, anus, anus, . . ."[18]

Athanasius Kircher was also interested in echo pranks. He describes being amused at a friend's expense in the Campagna at Rome, a lowland plain surrounding the city. His friend cried out, "*Quod tibi nomen?*" ("What is your name?"), and the echo impossibly replied, "Constantinus." The conceit was achieved using an accomplice hiding near a cliff where normally there was no echo. The accomplice would shout out the reply after hearing the question, impersonating the improbable sound reflection.[19]

A more impressive practical joker is Bob Perry, who has taught himself to impersonate an echo. He does an impressive performance of John F. Kennedy's inaugural address, complete with multiple renditions of each word as though caused by a public address system. With a little practice, you could teach yourself to do this. Select a spoken passage in which the space between syllables is a little longer than normal, as in JFK's speech, because of the slow delivery, and then say every syllable twice: "Ask ask not not what what you you can can do do . . .". To make it convincing, the echo word should be a bit quieter.

It is not usually the architecture that renders big speeches and announcements in train stations unintelligible; often electronics are to blame. Bad public address systems send out sound too loudly from too many sources. You hear words from two or more loudspeakers arriving separately because these sources of sound are different distances from you. One engineering solution is to change the position and orientation of each loudspeaker to make sure you hear only one at a time. Engineers can also use loudspeakers that illuminate defined areas rather than speakers that radiate in all directions—the aural equivalent of using a spotlight instead of a general-purpose lightbulb. But targeted illumination is not always possible, in which case engineers add electronic delays to each speaker to ensure that you hear all versions of the speech arriving at roughly the same time. Your brain will then lump the speech from the different speakers into one louder sound, minimizing the confusing cacophony of repetition.

On the TV show *Candid Camera*, echo prankster Bob Perry stood at Coit Tower, which affords great views of San Francisco, next to a false sign saying, "echo point." Standing alongside his unsuspecting victim, Bob shouted to create the illusion of sound bouncing off the tower, imitating an echo delayed by about one-fifth of a second. The joke was that whenever the victim tried shouting, there was no echo.

Bob Perry is impersonating what music producers would call a *slapback echo,* which is a single loud and delayed repetition. This effect was popularized in rock 'n' roll recordings from the 1950s and helped create the characteristic sound of famous singers like Elvis Presley. The audio engineers used two tape recorders to produce electronic echoes. A single big loop of the magnetic tape was fed through both recorders; the first machine would record the music onto the tape, and the second would pick up the sound from the tape a short time later, thus producing a delayed, slapback echo. The time between the passing of the tape under the record head on one machine to its reaching the pickup on the other machine determined the delay of the echo. On tracks such as "Boogie Disease" by Doctor Ross, the echo delay is about 0.15 second, creating the impression that the electric guitar on this blues recording is playing at double speed, as every strum is repeated.

The same effect gave Elvis's vocals a distinctive sound on his recordings with Sun Records, such as "Blue Moon." When Elvis switched to the RCA record label and achieved global hits with songs like "Heartbreak Hotel," the sound engineers could not work out how to reproduce the slapback echo, and they resorted to adding heavy reverberation from a hallway outside the studio.[20] Nowadays, it would be simple to reproduce this effect digitally, as delay is one of the cornerstones of modern pop production. To create the effect without electronics, RCA's engineers would have had to record Elvis in a studio next to a long tunnel or a tall room with a domed roof that had a slapback echo (remember, one of the dimensions would have to be at least 33 meters (110 feet), which would make it a rather large recording studio).

The Imam Mosque in Isfahan, Iran, might have worked for Elvis's voice as, according to the old writings on echoes, it is a *centrum phonocampticum,* the object of an echo. Constructed in the seventeenth century, the building is visually stunning, with dazzling blue Islamic tiles. A huge dome rises to an exterior height of 52 meters (170 feet) and, as one travel guide states, "replicat[es] individual sounds in a series of clear echoes."[21] Tour guides delight in standing underneath the dome and snapping or flicking a piece of paper, which creates a short, sharp "clack, clack, clack, . . ." The room immediately responds with about seven quick-fire echoes.[22] Sound bounces back and forth between the floor and ceiling, with the curved dome focusing the sound, forcing it to keep moving vertically up and down in a regimented fashion. Without a dome, the echo from the ceiling would be lost among all the other sound reflections in the mosque.

Luke Jerram is an artist who often uses sound as an art medium. His work *Aeolus* was inspired by a visit to Iran, where he heard the echoes in the Imam Mosque. I first met Luke about seven years ago, when we both reached the finals of FameLab, a pop idol–style competition to find science presenters for the media. I caught up with Luke when *Aeolus*, or what he called "my ten-ton musical instrument," was being installed outside my university's building at MediaCityUK in 2011.

Aeolus looks like a section through a giant steel hedgehog, an arch 4–5 meters (13–16 feet) high, with 300 long, hollow steel pipes sticking out of the top and sides (Figure 4.1). The shape was inspired by the twelve stuttering echoes that Luke had heard when he clicked his fingers in the mosque. Stand in the right place under *Aeolus*, and you can hear the focus from the arch subtly amplifying your voice. The light coming through the mirror-lined steel pipes creates geometric patterns echoing the decoration of the mosque.

While the arch is the most obvious visual feature of the sculpture, the main sound effect is created by long wires that stretch almost invisibly from supporting poles to the hedgehog. Each of the wires is driven to vibrate by the wind. Pieces of wood act like violin

Figure 4.1 *Aeolus.*

bridges, transferring the string vibrations to membranes stretched across the ends of the pipes. The membranes then cause the air in the pipes to resonate. Overall, the result is an eerie, pulsing sound like a minimalist piece of music by American composer Steve Reich in which tones come and go depending on the changing wind.

The work is named after the ruler of the four winds in Greek mythology. Luke's intention is to use "sound to paint pictures in people's imaginations," allowing visitors to "visualize the changing landscape of wind around the artwork."[23] The sound is hard to locate, appearing to be flowing vaguely from above. The lengths of the pipe are carefully selected to form a musical scale. Appropriately, the Aeolian mode is used, which, as a minor scale, lends a malevolent, spooky character to the sound.[24] If I shut my eyes, I could imagine I was in a B movie during a Martian invasion.

Luke decided to construct *Aeolus* after meeting a master digger in Iran who had described the construction of underground irrigation canals called qanats. The digging is wet, claustrophobic, and dangerous. The worst task is probably "devil digging," in which they mine up to a well of water from underneath. Just imagine being in a cramped passageway at the moment the digger breaks through and the water comes cascading down on top of you. What inspired Luke to create a singing building was the howling of a qanat's air vents in the wind.

Like the Iranian mosque, many grand buildings feature domes, but only rarely do they have the right curvature to achieve distinct echoes. The room diagrammed on the left in Figure 4.2 has a focal point that is too high; the building on the right brings amplified sound back to the listener at ground level and produces a pattern of repeated echoes. By measuring the time between echoes in a recording, I estimate that the interior height in the Imam Mosque is 36 meters (120 feet). The place to stand in the mosque is marked on the floor; this spot is called the *centrum phonicum*, according to old writings on echoes.

The surfaces of the floor and ceiling must also be made from materials that absorb little sound. The tiles of an Islamic mosque

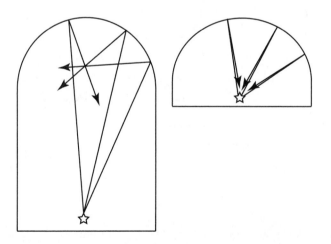

Figure 4.2 Focusing effects of two different rooms.

are ideal for two reasons: First, they are heavy, which means the sound wave is too weak to physically vibrate the tiles. Second, they are quite impervious to air, which means the acoustic wave cannot easily enter the tile and instead bounces off the front.

The Brixton Academy in London started life as the Astoria theatre, an art deco wonder from 1929. Film director Alfred Hitchcock was at the opening night, which featured Al Jolson's *The Singing Fool*.[25] The academy has an echo caused by reflections between the dome and the sloping floor,[26] but the echo is heard only during sound checks when the auditorium is empty. When the hall is packed, sound is absorbed by the audience as it squeezes into the pores of people's clothing, where the wave then loses energy. For poorly attended concerts, the echo has the fortunate effect of compensating for the small audience by amplifying the applause!

Brian Katz, a French academic and acoustic consultant, has been investigating, together with colleagues, an intriguing focusing ceiling from a long-lost room in Paris.[27] The room has been linked to the thousands of executions carried out during the French revolution. In the nineteenth century, Auguste Lepage wrote, "To this room [devoted] to meditation and prayer are linked bloody memories. It is there that was sitting the famous court . . . during the September 1792 massacres." Lepage goes on to describe the room: "Massive pillars supported a roof frame, which was a wonderful construction. This frame, rounded in a dome shape was made from Spanish chestnut, no nails were used and the thousand pieces that composed it were only fixed up by pegs."[28]

Though the room was destroyed in 1875, Brian is able to work with a nineteenth-century replica, a small-scale model conserved in the Musée des Arts et Métiers in Paris. The ceiling looks like an upturned wicker basket that has been squashed almost flat. Seen from below, the roof forms rings of beams with gaps between. While the curvature focuses sound, the focal point is at the wrong height for human ears to hear the effect. The secret behind the acoustic is that the beams are spaced farther apart near the middle of the ceiling, coming closer together at the edges of the room. Brian has shown that at some frequencies, the reflections from the different beams combine to amplify the sound in the middle of the room. It is a quirk of geometry. The dome's wooden lattice resembles a Fresnel zone plate, named after the French physicist Augustin-Jean Fresnel, who studied diffraction in the nineteenth century. Fresnel zone plates use diffraction to focus light. They can be used to focus laser beams and recently have been suggested as a lightweight substitute for heavy lenses used in space telescopes.[29] In acoustics, zone plates can be used to focus beams of ultrasound.

Echoes are not just fun phenomena; they can be aids to safety. A couple of years after the *Titanic* sank, a quick-witted captain described how his freighter had escaped a similar North Atlantic fate while sailing in foggy weather off the Grand Banks of Newfoundland. A five-second blast on his ship's foghorn resounded from the mist. But was this another steamer's call? The captain sounded a more elaborate pattern of blasts, which were exactly repeated back to him, revealing that this was an echo. The *Day* newspaper described how he took evasive action to "avoid hitting a berg that he could feel and hear, but could not see."[30]

Another historical example of sailors using echolocation comes from Puget Sound in Washington State. An article in *Popular Mechanics* magazine from 1927 describes the inside passage from Puget Sound to Alaska as "more crooked than the famous dog's hind leg; a narrow tortuous channel."[31] In foggy weather, navigators would locate themselves by listening to the echo of their steamboat's whistle. The strong tides in the channels made it impossible for the ship to move slowly, as would be done in foggy conditions out in open sea. As the same magazine article explained, "The full steam ahead—followed by the full speed astern—is the rule of the echo pilots."[32] If an echo was delayed by 1 second, the whistle would have traveled 340 meters (370 yards), indicating that the boat was 170 meters (185 yards) from shore. Sailors learning to navigate the route had to memorize the reflection delays from key landmarks. On a small island that was too low to produce an echo, an 8-meter-square (25-foot) signboard was erected to aid navigation by reflecting a blast on the whistle.

The magazine claims that the pilots identified the type of coastline from the echo: "A low coastline returns a 'sizzling' echo while a high cliff gives a solid 'plunk.' The echo 'scrapes' from a beach of sand or gravel, and a forked headland yields a double echo to the expert ear."[33] This claim strained credulity until I heard a talk by Norwegian acoustic expert Tor Halmrast, who had been carrying out experiments into echolocation by blind people.

By making a clicking sound and listening to the sound of the reflections, people can learn to navigate with their ears, mimicking the techniques used by dolphins, bats, and oilbirds. Daniel Kish learned to echolocate from a young age, and in *New Scientist* he described what a hectic school day was like for him at age 6:

> Clicking my tongue quickly and scanning with my head, I move cautiously forward . . . the sounds in front of me take on a softer hue, suggesting there's a big field of grass ahead . . . Suddenly, there is something in front of me. I stop. "Hi," I venture, thinking at first that someone is standing there quietly. But as I click and scan with my head, I work out that the something is too thin to be a person.
>
> I realise it is a pole before I reach out to touch it . . . There are nine poles all in a line. I later learn that this is a slalom course, and while I never attempted to run it, I practised my biking skills by slaloming among rows of trees, clicking madly.[34]

The clicking is usually made by sharply dropping the tongue in the mouth, maybe with an accompanying suck or a short, sharp cluck. The exact sound is very individual, making it difficult for one person to echolocate using someone else's vocalization.[35] The diversity of sounds one can make is amazing. A palatal click is a good one, made by a quick release of the vacuum produced between the tip of the tongue and the roof of the mouth. Being short and loud, it is easier to pick out in noisy places.

The palatal click also contains sound spread across many frequencies, which human echolocators find useful.[36] Since they are listening for surfaces only a few yards away, and since most reflections from such surfaces arrive too quickly to be heard distinctly, these people must learn to detect subtle changes between what is heard by each ear. The interference of the click and its reflection might cause coloration (a change in the frequency balance), altering the tonal quality, or what musicians call *timbre*. The reflection might elongate the original click, for example, suggesting a reflection from a nearby surface. The effect depends on the distance from the reflecting surface, which alters the delay, and also on how the object reflects acoustic waves: larger objects reflect low frequencies more strongly; soft objects tend to absorb sound and so produce weaker reflections. Studies show that even novice echolocators can learn to distinguish square, triangular, and circular shapes with minimal practice.[37]

Some of the most extraordinary echoes come from man-made structures. Engineered curves can focus echoes, and parallel flat walls can encourage sound to bounce back and forth in a way that is unlikely to happen with natural surfaces. Bridge arches have great potential to be sonic wonders, as I found out on a canoeing trip on the Dordogne River in France a couple of months before seeing *Aeolus*. One stone arch was just the right size and shape to put the focal point at the water level, and slapping my paddle on the water created a wonderful ricocheting sound. During the lunch break I explored under another bridge over a sandbank. Standing with my back to the edge of the arch and clapping my hands produced a stunning fluttering sound—a multiple echo.

Across the Atlantic in Newton Upper Falls in Massachusetts, a similar fluttering sound under an aqueduct proved so remarkable that denizens named it Echo Bridge. Built in the 1870s, this 40-meter-wide (130-foot) arch spans the Charles River and even has steps down to a specially built platform so that visitors can test out the sound effect. The Internet has several videos of dogs being driven mad by their own echoes, believing there to be a rival canine on the other side of the river. Not only does the bridge attract tourists and playful dog owners; it also intrigues scientists. In September 1948, Arthur Taber Jones wrote to the *Journal of the Acoustical Society of America*, detailing a small study. "A handlcapp [sic] is returned in a series of about a dozen echoes of decreasing loudness, and at a rate of about four echoes per second."[38] Jones describes elaborate experiments undertaken to determine what is causing the reflection.

The question Jones was trying to answer was whether the sound was skimming around the inside of the curved arch, like the whispering galleries I describe in the next chapter, or propagating horizontally just above the water. He unsuccessfully tried using listening trumpets to determine which direction the sound was coming from. Further attempts using blankets to block off sound going around the arch failed because of high winds.

Unable to get to the bridge myself, I found photos and postcards that allowed me to estimate the shape of the archway. I calculated the echo delay from video soundtracks of dogs barking. And finally, modern prediction methods allowed me to visualize how the sound moves.

Figure 4.3 shows twelve frames from an animation I made to

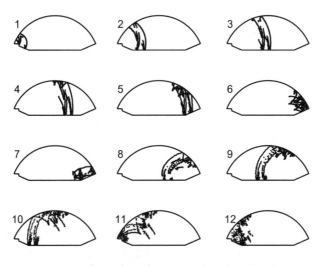

Figure 4.3 Snapshots from an animation showing
sound moving under Echo Bridge.

understand the bridge. Each snapshot shows the roughly semi-circular shape below the arch, where the platform is to the left and the water is the long flat line at the bottom. Starting from the top-left frame, the dots show how sound moves from the talker, across to the far side of the bridge, and back again.

To make this animation, I modeled sound as lots of tiny snooker balls, which are fired in all directions from the platform. The computer works out how the balls bounce around the odd-shaped billiard. For images 1–6 in Figure 4.3, the sound is moving from left to right; it then reflects from the right side and travels back in the opposite direction. The answer to Jones's question is that the sound both hugs the inside of the curve and skims along the water's surface.

Early writers on echoes were keen to find extraordinary multiple echoes with the greatest numbers of repeats—echoes that would turn a "ha" into laughter. This endeavor was taken to the absurd by the echo collector in Mark Twain's short story "The Canvasser's Tale":

> You may know, sir, that in the echo market the scale of prices is cumulative, like the carat-scale in diamonds; in fact, the same phraseology is used. A single-carat echo is worth but ten dollars over and above the value of the land it is on; a two-carat or double-barreled echo is worth thirty dollars; a five-carat is worth nine hundred and fifty; a ten-carat is worth thirteen thousand. My uncle's Oregon-echo, which he called the Great Pitt Echo, was a twenty-two carat gem, and cost two hundred and sixteen thousand dollars—they threw the land in.[39]

In the seventeenth century, the real-life collector and myth buster Marin Mersenne analyzed the claim that a tower near the Aventine Hill in Rome would repeat the entire first line of Virgil's *Aeneid* eight times.[40] Since it takes nearly 40 seconds to hear eight repeats of the phrase, the farthest reflection would have had to travel a round-trip of 14 kilometers (8½ miles), which is too far for the voice to carry and still be audible.

More believable are the stories about the sixteenth-century Villa Simonetta in Milan. The great eighteenth-century mathematician Daniel Bernoulli stated that he could hear up to sixty repetitions from the echo.[41] Twain wrote about this in his travel book *The Innocents Abroad*, which includes a plate showing a woman entertaining two gentlemen by blasting a trumpet to excite the echo. Iris Lauterbach, writing on Italian gardens, noted that the villa was famous well into the nineteenth century "but not for its garden: the attraction was an echo."[42]

The villa was a rectangular horseshoe shape, with two large wings, exactly parallel to each other, spaced 34 meters (110 feet) apart. The semi-enclosed courtyard used to open up to a luxuriant garden. On the first floor there was a single window up near the roof on one of the wings. Speak from this window and the words would bounce back and forth across the courtyard between the parallel wings. The sound took 0.2 second to complete the round-trip, which meant a very short blast repeated many times. Old reports claim that a pistol shot would repeat forty to sixty times.[43] Seventeenth-century engravings of the villa show the upper walls of the wings to be very simple flat surfaces, ensuring that the sound could bounce back and forth without being scattered in other directions and lost from the echo path.

In the engravings, the echo window looks odd—the only opening on the upper walls of the wings, and ruining the architectural symmetry. It makes me wonder whether the window was deliberately placed to take advantage of the acoustic phenomena. Unfortunately, the villa was extensively damaged by bombing during the Second World War, so the courtyard now lacks the grand colonnades and vistas, and disappointingly, the echo has been dulled to a single retort.[44]

Is it just me, or is it virtually impossible not to shout and whoop when entering a tunnel? Some are better than others, with the foot tunnel under the Thames near Greenwich, London, being one of my favorites. Finished in 1902, it was built to let South London residents walk to work across the river in the Isle of Dogs. I went back there a few months after my trip to France, on a cold winter's night, to see if my childhood memories of the acoustics were right. Despite being a foot tunnel, nearly everyone passing through seemed to be on a bike. I spent some time wandering up and down the 370-meter-long (400-yard) tube. It is a squat cylinder covered in off-white glazed tiles.

The poorly lit tunnel is only about 3 meters (10 feet) in diameter. As sound waves bounce back and forth across the width, they tend to distort dramatically. If I stood right in the middle, my voice resounded with a metallic twang. The resonances of the tunnel were overamplifying certain frequencies in my voice, making it sound unnatural. I asked the sound artist Peter Cusack about his impressions of the place:

> Sometimes there is a busker in the middle of it, and if you lis-ten from the end . . . then there's no way of telling what the tune is or even what instrument they are playing. It is just this musical mush that comes up one end, which is actually quite pleasant. And as you walk down the tunnel and get closer and closer, and it becomes clearer and clearer, it is often a bit of a disappointment when you get there.[45]

At one point I was alarmed to hear what I initially thought was an approaching freight train. I was relieved to see it was just the rumble of a skateboard being amplified by the tunnel. After passing me, the skateboarder flipped up his board but missed catching it, causing an impressive clash as though someone were slamming the doors of a large cathedral. The initial crash traveled hundreds of yards to the end wall and returned with an audible echo. The hard-tiled surfaces allow the sound to rattle around the tunnel for a long time before dying away.

Engineers at Bradford University, England, have been using the ability of tunnels to carry sound long distances to find obstruc-tions in sewers. Noise is played down the pipe, and a microphone records any echoes heard. The time it takes for the echo to arrive reveals how far away the blockage is, and the acoustic character-istics of the reflection tell the scientists about the size and type of obstruction.

One reason tunnels often have impressive acoustics is that sound can travel an unusually long way in them. If someone is talking to you outdoors, the farther apart the two of you are, the quieter the other person sounds. Imagine blowing up a balloon: as it expands, the rubber gets thinner as it spreads over a larger surface area. Being farther from a sound source outdoors is like being on the edge of the balloon; the energy is spread thinly like the balloon's rubber, so it is quieter. But in a tunnel the acoustic wave is spread across the width of the tube, which does not change in size as you get farther away from the sound source. The only way energy is lost is through absorption by the tunnel walls. If the walls are made of hard materials such as tiles, brick, or sealed concrete, sound can carry for huge distances.

Still curious about why my voice sounded so metallic in Greenwich, I sought out another example to experience—one where the effect is even stronger. London's Science Museum has a hands-on gallery full of children noisily enjoying science. Across the back wall is a long, sloping industrial tube, 30 meters (100 feet) long and about 30 centimeters (12 inches) in diameter. "Sounds like gunfire," suggested a young boy just before I started to experiment. This was a good description. A clap of my hands sounded like a cross between a sheet of metal being struck and a laser gun from a sci-fi movie with a slow recoil.

It is easy to assume that the tube's material dominates the sound. But while the pipe was made of metal, the material has very little to do with why my voice or clapping developed a robotic quality. The tube could have been made of any hard material, such as concrete, metal, or plastic, and it would still have made a twang, as happened within the tiled Greenwich foot tunnel. What is most important is the geometry of the bore, because it is the air that is doing most of the vibrating, not the tube walls. The same confusion exists with musical instruments. I learned the clarinet when I was younger, and the lower notes on the instrument are often described as being distinctly "woody," which you might assume comes from the black ebonite tubing. However, my colleague Mark Avis once played a brass clarinet and noticed how remarkably "woody" that sounds. The great jazz player Charlie Parker famously used a plastic saxophone for some gigs, yet he still created his distinctive sound.[46]

Similarly, the "brassy" blast of a trumpet or trombone might be wrongly attributed to the metal from which it is usually made. Some historic brass instruments, such as the cornetto, were actually made of wood and yet can still make a "brassy" sound. A musical instrument simultaneously generates many different frequencies, known as harmonics, which give distinct color to the sound. When an oboe plays a tuning note for the orchestra—a concert A at 440 hertz—sound is also produced at 880, 1,320, and 1,760 hertz. These *harmonics* are multiples of the *fundamental* frequency, and their strength depends on the instrument's geometry. When a trombone is played loudly, a shock wave can be created inside the bore similar to that produced by a sonic boom, generating lots of high frequencies. A "brassy" sound is associated with musical notes that have exceptionally strong high frequencies.

The echo tube at the Science Museum has only a few strong harmonics, and these are not simple multiples of the fundamental. Musical instruments sound beautiful because they have been designed to produce harmonics whose frequencies are regularly spaced. Large pieces of metal tend to radiate at irregular frequencies and sound dissonant. Thus the tube, with its discordant frequencies, adds a metallic quality to voices. Another key feature determining a musical instrument's voice is how notes begin and finish. A metal chime bar can ring beautifully for a long time; similarly, the air in the echo tube at the Science Museum rang on and on.

But something else intrigued me about the echo tube: clapping my hands created a zinging sound, the echo starting at a high frequency and then descending in pitch. I talked to a few colleagues,

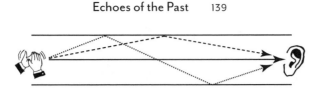

Figure 4.4 A single hand clap at one end of a long tube
and listening at the other end.

and they were similarly bemused because none of us expected a shift in frequency in a simple tube. One of the fun things about being a scientist is having your expectations subverted and finding something new to understand. Looking through the literature, I found that the descending zing was a *culvert whistler*. It was first documented a few decades ago by the late American scientist Frank Crawford, who observed a chirp from a pipe under a sand dune in California. In an effort to explain his observation, reported one article, "Crawford has clapped his hands, beat bongo drums and has banged on pieces of plywood in front of culverts all around the San Francisco Bay Area."[47]

If you listen to one end of a culvert while someone claps hands once at the other end, as illustrated in Figure 4.4, the first sound to arrive travels straight down the middle of the tube following the shortest distance. The next sound to arrive has reflected once off the side wall and so has traveled a little farther. The next sound has hit both sides once while zigzagging down the tube. Later sounds will follow a longer, more jagged path. If you plot these sounds arriving over time, as illustrated in Figure 4.5, you find that the reflections arrive close together at first, and then gradually are spaced farther apart toward the end of the chirp. At any particular instance, the pitch of the chirp is determined by the spacing between adjacent reflections. When reflections arrive rapidly one after another, as happens initially, a high-frequency sound is the result. As the time between reflections increases, the frequency lowers.[48] A similar

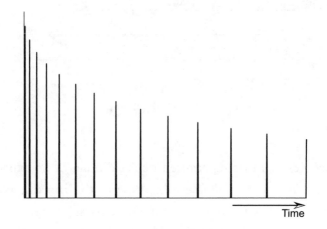

Figure 4.5 A clap and its reflections from inside a culvert. (Each clap is simplified to a single peak so that the pattern of claps arriving is clearer.)

downward glissando also happens when vibrations pass through a solid like a metal. This might be another reason why the echo tube sounds metallic.

Multiple reflections lie at the heart of echoes that create almost-musical sounds. Not long after my canoeing trip, on a hot, sunny afternoon in the city of Angoulême in France, I stood outside the comic book museum while inside my children devoured the extensive *Asterix* and *Tintin* collections. Bored, I experimented with clapping and listening to the reflection from the front of the building, a wide, low, and white converted warehouse that had been used to store cognac. But it was the reflection from another structure that caught my attention. There was a high-pitched sound to my right, like someone squeezing a squeaky toy, coming from a staircase. A *tonical* echo! Boredom turned to an afternoon of fevered experimentation as I recorded and documented the strange reflection from this short flight of stairs.

What I was hearing was the same phenomenon that creates the chirping Mayan pyramids described in Chapter 2. Staircases can make many different sounds. Acoustic engineer Nico Declercq wrote to me about a quacking staircase: "It is on the Menik Ganga (Gem River) in Sri Lanka, a river you must cross in order to reach the sanctuary of Katharagama ... [W]hen you cross the water ... you can hear quacking ducks when you clap your hands or when women hit rocks with clothes they are washing."[49] Back in Europe, artist Davide Tidoni popped balloons to reveal the unusual acoustics of the Austrian city of Linz, including an explosive wheezing sound created by a very long staircase.[50]

The strange sounds are made by the pattern of reflections from the treads of the stairs, which distort the sound of the balloon pop or clap, and this pattern can be explained by geometry (Figure 4.6). Figure 4.7 shows the ninety reflections, one from each stair tread, that arrive if you clap your hands once in front of the Mayan pyramid El Castillo. The frequency drops by about an octave because the spacing between the reflections roughly doubles.

Probably the best way of analyzing a chirp is to look at the

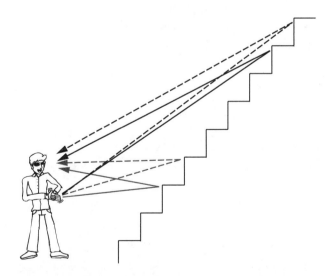

Figure 4.6 Sound chirping from a staircase.

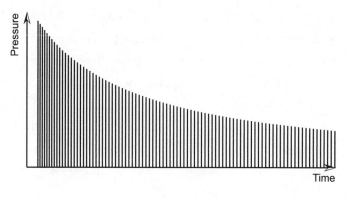

Figure 4.7 Reflections of a single clap from the staircase of El Castillo,
the Mayan Temple of Kukulkan.

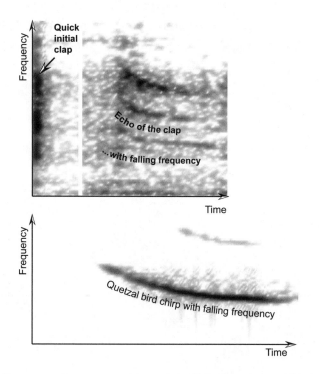

Figure 4.8 Acoustic signature of the Kukulkan pyramid (top)
and a quetzal bird (bottom). (The echo has been amplified
so that the drooping lines of the chirp are easier to see.)

spectrogram, as I used previously with bat calls. The top image in Figure 4.8 shows the chirping echo from the staircase. The black, vertical line at far left represents the initial clap. The fuzzy, dark lines that droop to the right show the reflections in which the pitch is decreasing. Compare this sonic fingerprint to the cry of the quetzal bird, the bottom image, which features a similar drooping line. This similar decrease in pitch explains why some people believe the staircase echo resembles a chirping bird.

The particular sound reflected from a staircase depends on where the clapper stands, as well as on the size and number of steps. The squeaky stairs outside the comic book museum were quite short and did not have enough reflections to create the extended sound of a chirping bird. The longest staircase in the world runs alongside the funicular railway up the Niesen, a mountain in Switzerland. It is opened to the public only once a year, for a marathon, and the winner takes about an hour to climb the 11,674 steps. When I simulated the staircase in an acoustic model, it sounded like a wheezy air horn.

If you're looking for a staircase to experiment on, I would suggest finding one in a quiet place away from other reflecting surfaces. It does not have to be very long, maybe twenty steps, but the more stair treads there are, the more impressive the effect will be.

Archaeologists argue about the role of staircases on the sides of Mayan pyramids, and whether they were built to imitate the chirp of a quetzal bird. Leaving aside this debate, what other sounds could the Mayans have made if they had built the stairs differently?

The sound reflected from a flight of stairs is determined by the pattern of reflections that build up as a clap bounces off each stair tread and returns to the listener. In a normal staircase, the later reflections arrive farther apart than the earlier ones, causing a chirp that descends in frequency. Imagine a staircase constructed by bad builders—one in which the steps are not all the same size.

At the bottom of the stairs the steps get smaller and smaller as they go up, creating a series of reflections that are heard with a rising pitch. Then, toward the top, the steps stretch out and get bigger and bigger to create a quick drop in the pitch. With just the right pattern of steps between about 3 and 10 centimeters (1–4 inches) wide, you can get a chirp that rises and then falls in frequency; in other words the staircase would make a wolf whistle. A completely useless staircase, but what a sonic wonder it would be!

While the embellishment of my voice by a tunnel was not pleasant, it explains why old writings on tonical echoes observe voices modulated into distinct tones. Clapping near a staircase shows how reflections outdoors can sound like a distinct musical note. Occasionally, the old echo tales are fanciful, with the most unlikely one featuring a tune being played on trumpet that resounds at a lower pitch.[51] A change in pitch flouts the laws of physics, but then so does the phrase "a duck's quack doesn't echo," and people seem happy to keep repeating that. Maybe the trumpet echo was a simple practical joke, or maybe the basis is a more subtle tonal coloration that has just been overembellished as the story is retold.

No matter how powerful an echo is, or what type it may be, all the echoes described in this chapter have one thing in common: they can be enjoyed with just one ear; that is, they are monaural delights. Let's turn now to binaural sonic wonders—those that mess with how our brains use two ears to localize sound.

5

Going round the Bend

Whispers reflected from a giant hemispherical ceiling were described by Wallace Sabine, the grandfather of architectural acoustics, as "the effect of an invisible and mocking presence."[1] In the huge dome of the Gol Gumbaz mausoleum in India, "the footfall of a single individual is enough to wake the sounds as of a company of persons," reported the celebrated physicist C. V. Raman, and "a single loud clap is distinctly echoed ten times."[2] When I was in the sewer (see the Prologue), my speech appeared to hug the walls of the tunnel, spiraling around the inside of the curve as the sound slowly died away. Some of the strangest sound effects can be created by simple concave surfaces.

In 1824, naval officer Edward Boid described how a curve can dramatically amplify sound, and not always for the best. He wrote, "In the Cathedral of Girgenti, in Sicily, the slightest whisper is borne with perfect distinctness from the great western door to the cornice behind the high altar—a distance of two hundred and fifty feet." Unfortunately, the confessional was badly sited: "Secrets never intended for the public ear thus became known, to the dismay of

Figure 5.1   The cat piano.

the confessors, and the scandal of the people . . . till at length, one
listener having had his curiosity somewhat over-gratified by hearing his wife's avowal of her own infidelity, this tell-tale peculiarity
became generally known, and the confessional was removed."[3]

For centuries, people have known that curved surfaces amplify
sounds and allow covert listening. Athanasius Kircher gave a good
explanation in the seventeenth century. We met Kircher in Chapter 4 because he wrote extensively on echoes. His publications also
document some fantastical devices, including giant ear trumpets
built into the walls of royal chambers for eavesdropping. Probably his most famous—or infamous—device is the *Katzenklavier*
(literally, "cat piano"; Figure 5.1). It has a normal piano keyboard
in front of a line of cages, each of which has a cat trapped inside.
Every time a piano key is pressed, a nail is driven into the tail of one
unfortunate feline, which naturally screeches. With the right set of
cats, ones that shriek at different frequencies, a sadistic musician
could play a tune on the instrument. The sound would have been

excruciating, but then it was designed to shock psychiatric patients into changing their behavior, rather than being a genuine instrument for playing Monteverdi or Purcell. Fortunately, it is unlikely that it was ever built.

At this point you might be doubting the sanity and rationality of Kircher. Yet he drew diagrams that illustrated a good scientific understanding of how an elliptical ceiling can enhance communication between two people (Figure 5.2).

The lines in the diagram show the paths that sound "rays" take when going from the speaker to the listener. These ray paths can be worked out using a ruler and protractor. Alternatively, by treating the room as a weird-shaped pool table, the paths can be worked out by following the line a cue ball would take (ignoring gravity). If the

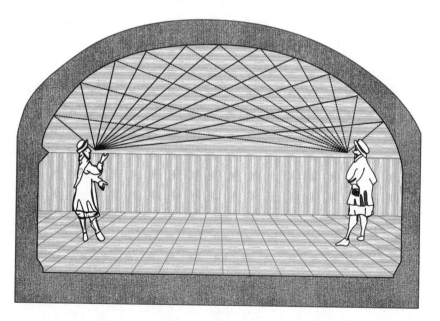

Figure 5.2 Simplified tracing of a plate from Athanasius Kircher's
Phonurgia Nova (1673).

cue ball is placed at the speaker's mouth and fired toward the ceiling, it will always go to the listener. So all the sound going upward is focused at the listener, allowing even quiet whispers to be heard across a large room.

The problem with this design is that the listener and speaker have to stand in particular places—the foci of the ceiling ellipse. The design is not very useful if one person wants to talk to an audience of listeners scattered around the room. In 1935, the Finnish modernist architect Alvar Aalto tried to overcome this problem using a wavy ceiling for the Viipuri Library. (Originally, the library was in Finland, but the town of Viipuri was subsumed into the Soviet Union after the Second World War.) From the speaker's podium at one end of the room, the ceiling looks like gentle undulating waves coming in from the sea. The wave troughs form concave curves, each designed to amplify the sound for particular listeners. Unfortunately, each wave crest also reflects sound back toward the talker, weakening the strength of the reflections to the back of the room and making it harder for those at the rear to hear the speaker. In reality, using curved focusing ceilings to improve communication in a room rarely works as intended.[4]

Elliptical ceilings work rather like a shaving mirror, a simple curved reflective surface that brings light rays together to a point. Both the ceiling and shaving mirror achieve magnification, but whereas for light the result is a bigger image, for sound the result is increased loudness. In the shaving mirror, the reflections that meet your eyes are distorted so that you see an enlarged picture of your face. But with hearing, the reflections coming from different parts of the ceiling add together at the entrance of the ear canals and are treated holistically by the brain. The overall effect is a louder sound, which can make distant objects appear closer than they really are.

In *Elements of Physics* (1827), Neil Arnott writes:

The widespread sail of a ship, rendered concave by a gentle breeze, is a good collector of sound. It happened once on board a ship sailing along the coast of Brazil, far out of sight of land, that the persons walking on deck, when passing a particular spot, always heard very distinctly the sound of bells, varying as in human rejoicings. All on board came to listen and were convinced; but the phenomenon was most mysterious. Months afterwards it was ascertained, that, at the time of observation the bells of the city of Salvador, on the Brazilian coast, had been ringing on the occasion of a festival; their sound, there-fore, favored by a gentle wind, had traveled 100 miles [160 kilometers] by smooth water, and had been brought to a focus by the sail on the particular spot where it was listened to.[5]

Is this story true? Can an acoustic mirror pick up bells 100 miles away? One way to answer this question is to look at some more modern examples. Just south of Manchester in England stands the gigantic dish of the Lovell Telescope at Jodrell Bank Observatory. This telescope uses the same process of focusing to collect and magnify radio waves, and in the past it played an important role in the space race. When the Soviet probe *Luna 9* surprised the West by landing on the moon in 1966, the observatory intercepted the spacecraft's transmissions. Feeding the signal into a fax machine revealed pictures of the lunar surface that were then first published in a British newspaper before they appeared in the Soviet Union.

Two whispering dishes stand in the shadow of the giant tele-scope. (There are other, similar whispering dishes at other science museums and sculpture parks.) The last time I visited, my teenage sons entertained themselves by whispering insults at each other using the dishes. The mirrors are 25 meters (80 feet) apart, yet the sniping siblings were very loud. But Arnott's sailing ship was much farther from Salvador than a few tens of meters.

Around the coast of England are remnants of acoustic mirrors designed to work over relatively long distances. These are large, ugly concrete bowls, typically 4–5 meters (13–16 feet) in diameter, which face the sea. Built in the early twentieth century, they were intended as an early-warning system for enemy aircraft. Most are bowl-shaped, but in Denge, Kent, there is also a vast, sweeping arc of discolored concrete. The arc is 5 meters (16 feet) high and 60 meters (200 feet) wide—the equivalent of about five double-decker buses parked end to end. It is curved both horizontally and vertically to magnify the engine noise from approaching aircraft.

Military tests showed that the large strip mirror could detect aircraft 32 kilometers (20 miles) away, when enemy planes were roughly a third of the way across the English Channel. But in poor weather conditions, aircraft might get within 10 kilometers (6 miles) before detection, and listeners struggled to hear planes with quieter engines.[6] Even on a good day, these acoustic mirrors provided a measly ten minutes of extra early warning. Once a working radar system was developed in 1937, the plan to build an extensive network of mirrors was dropped.

The short detection ranges of the concrete acoustic mirrors makes the claim of a ship sail focusing sounds from a festival 100 miles away seem fanciful. But a catastrophic event several years ago in England hints at an explanation.

In December 2005 an overflowing storage tank caused a giant explosion at the Buncefield oil terminal in the UK and shook glass doors in Belgium 270 kilometers (170 miles) away.[7] This was one of the biggest explosions in peacetime Europe, measuring 2.4 on the Richter scale.[8] Although the bang at Buncefield must have been very powerful, the initial loudness alone does not explain the huge distances the noise carried.

The catastrophe happened on a still, clear, and frosty morning when a layer of cold air was trapped close to the ground by warm air above. Without this temperature inversion, the Belgians would have been left undisturbed. When the oil refinery blew up, sound waves would have been sent out in all directions, rather like the ripples created when a rock is lobbed into a pond. Much of the noise would have headed upward toward the heavens and, under normal conditions, would never have been heard again. But with the temperature inversion, the sound heading upward was refracted back down to Earth and could be heard far away.

Intriguingly, Arnott's story of the sailing ship mentions the weather as being a crucial part of the tale. The report might well be correct if a temperature inversion helped direct sound to the concave sail.

A few years ago I presented two science shows at the Royal Albert Hall in London to thousands of children. Though better known as a music venue, the hall is actually dedicated to the promotion of art and science, and it was built on land purchased with the profits of the Great Exhibition of 1851. For an amateur performer like me, a complex science show is a daunting challenge, made all the worse in this case by the vastness of the arena. Fortunately, the acoustics have been significantly improved since the hall opened 130 years ago. Indeed, the Prince of Wales struggled with his opening speech. According to the *Times* (1871):

> The address was slowly and distinctly read by his royal Highness, but the reading was somewhat marred by an echo which seemed to be suddenly awoke from the organ or picture gallery, and repeated the words with a mocking emphasis which at another time would have been amusing.[9]

The hall's ubiquitous curved surfaces are probably what caused the mocking echoes. From above, the floor plan appears as an ellipse, and the whole structure is topped with a large dome. The curved surfaces focus sound like Kircher's elliptical ceiling. But how such reflections are perceived depends on the size of the room. In the vast Royal Albert Hall, the curves cause disastrous echoes. Sound appears to come from several places in the room and not just the stage. In a small room the focused sound arrives quickly; in a larger room the reflections are delayed.

You can test this phenomenon out with a friend.[10] Find a large open area with one big reflecting wall, such as a large building at the side of a park, or the side of a quarry. A quiet place away from noise is best. For the exercise to work, you need to hear the sound bouncing off the wall with few reflections from other surfaces. The wall does not need to be curved if it is large enough. If you and your friend stand some distance apart but each of you is the same distance from the wall, then the effect is more obvious. The test would work particularly well on a snowy day, when the sound reflecting from the ground would be absorbed by the snow and all traffic would have ground to a halt.

Walk toward the wall while chatting with your friend, and at some point you will become aware of the reflection from the building. As you get closer, you will hear the echo from the wall becoming louder because the reflection has traveled a shorter distance. But then as you approach even nearer, starting about 17 meters (56 feet) away, the sound of the reflection will appear to gradually get quieter until, at about 8 meters (26 feet) from the wall, it appears to disappear altogether. It is still there, but it is no longer being heard separately: your brain has combined it with the sound traveling directly in a straight line from your friend.

The way the brain combines sounds is important, because otherwise we would rapidly become overwhelmed by the vast number of reflections that accompany us. As I type this sentence, the rattle of the keyboard is being reflected off the desk, the computer monitor, my phone, the ceiling, and so on. Yet my hearing is not over-whelmed by all these different reflections; the sound still appears to be coming straight from the keyboard as it should be.

The same thing happens in Kircher's small room. The reflections from the elliptical ceiling arrive quite quickly, and unless the reflections are very loud, the brain does not hear them as separate from sound traveling directly between talker and listener. By contrast, the Royal Albert Hall is so vast that the focused reflections arrive much later, creating "mocking" echoes.

Acoustic engineers made various attempts to remove the echoes from the Royal Albert Hall. The most successful solution was to add the "mushrooms" hanging from the ceiling. Originally installed in 1968 following an idea by Ken Shearer of the BBC, these large discs hang down at the base of the dome and reflect sound away from it.

Although it is no longer possible to appreciate (or not) the echo from the hall's ceiling, there are plenty of other domes to explore. A few miles from my home, for example, is Manchester Central Library, which has a great dome with a focal point that used to be close to the microfiche machines. Every time a glass plate fell to cover the microfiche, a startlingly loud echo resounded from the ceiling.

The library is currently closed for renovation. One hopes that the building work will not be as sonically unsympathetic as the renovation of the US Capitol in the nineteenth century in Washington, DC, which ruined a fine ceiling echo from a famous whispering dome.[11] The Capitol's dome used to be an almost perfect hemisphere centered at the head height of visitors, and although the ceiling appeared coffered with indented squares, it was actually smooth, with *trompe l'oeil* painting creating the illusion of structure and texture. Before 1901, this domed space was a great draw for tourists. According to the *New York Times* in 1894:

The whispering gallery still holds the palm among the show places of the great marble structure. Once in a while an old resident of Washington is initiated into the mysteries of the echoes and other acoustic phenomena which abound in this old-time chamber, and he feels a little ashamed of his tardiness in seeking this remarkable entertainment.[12]

But while it was great fun for the tourists, it was a poor place for the House of Representatives to hold debates. As the *Lewiston Daily Sun* put it in 1893:

> The orator who was not cautious enough to remain in one spot during the delivery of his address found the acoustics of the hall taking strange liberties with his elocution, transforming his crescendo sentences into comical squeals, or causing his pianissimo phrases, his stage whispers, to shriek and wail as he moved to and fro from one echo point to another.[13]

A gas explosion and fire in 1898 elsewhere in the building led to replacement of the wooden dome by a fireproof construction. Real plaster coffers were installed in place of the *trompe l'oeil*, making the effect of the focus weaker and less remarkable. As the eminent acoustician Lothar Cremer remarked, "To everyone's dismay, the famous focusing effect was remarkably reduced, since the precise geometric reflection had been supplanted by a fuzzier diffuse reflection."[14]

Changing from a smooth surface to one covered in lumps and bumps is rather like taking a perfect optical mirror and scratching it badly or adding a frosting. The irregularities of the surface cause the light or sound to be scattered away from the focal point. For an optical mirror, the result is a fuzzy image; for the Capitol's dome this scattering weakens the sound reflections, so whispers are no longer remarkably loud and voices are less distorted.

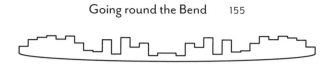

Figure 5.3 Diffuser designed for the curved wall of the
National Museum of the American Indian.

The effect of the coffers on the Capitol's focus reminds me of an engineering project I worked on a few years ago. I designed diffusing surfaces for the large circular Rasmuson Theater at the National Museum of the American Indian in Washington, DC. To keep the curves from causing focusing and creating mocking echoes, I designed a bumpy surface that, like the coffers in the Capitol's dome, scattered sound in all directions away from the focal points. A cross section through the diffuser looked like a city skyline (Figure 5.3). As sound waves meet the diffuser, the irregular heights of the blocks force the reflections to go off in different directions.

My innovation was coming up with a way of determining where to place the "skyscrapers" and how high to make them. I used a trial-and-error process in which a computer program tries out many different skylines. For each configuration, the program predicts how sound will reflect from the surface and evaluates whether the focus for the curved surface will be removed. The program keeps rearranging the skyscrapers until a good design is found. This iterative process, known as *numerical optimization*, has been used in many branches of engineering, including designing parts for the space shuttle. One of the reasons this method is so powerful for acoustic diffusers is that it allows surfaces to be designed that fit the visual appearance of the room. The acoustic treatments do not have to look like ugly add-ons. Curves, skylines, pyramids— whatever shape the architect desires—the method can be tasked with finding features that have the best acoustic performance.[15]

> The great thing about a dome is that if you stand right under
> the centre and clap your hands, you will be deafened a moment
> later by the echo of your hands clapping. If you say, in mock
> horror, "A HANDbag?", you will hear the echo of Dame
> Edith Evans a second later, from the heavens.[16]

This is the journalist Miles Kington encouraging you to unleash
your inner Lady Bracknell from *The Importance of Being Earnest*
by Oscar Wilde. But while domes are fun, a completely spherical
room is even better because reflections are amplified even more.

The Mapparium in Boston is a 9-meter-diameter (30-foot) sphere
and was built in 1935 following a suggestion by architect Chester
Lindsay Churchill. It is a giant hollow globe of the world, with the
seas and continents vividly drawn on stained glass. It took eight
months to paint and bake all 608 glass panels, which are mounted
on a spherical bronze frame. Visitors traverse a walkway cutting
through the center of the Earth linking up two opposite points
on the equator. Three hundred lightbulbs illuminate the globe from
the outside. Looking at the world from the inside out is an odd
experience, but what also strikes visitors are the strange acoustics,
which were an accidental by-product of the geometry.

William Hartmann, from Michigan State University, and colleagues documented the various illusions that can be heard. Usually,
moving farther away from a listener makes a talker's voice quieter,
but that's not always the case in a spherical room. Imagine, Hartmann writes, "you are on the Mapparium bridge two meters [about
7 feet] to the left of dead center. Your friend is exactly at the center
and is talking to you. His voice seems rather quiet. Now your friend
walks away from you, and his voice gets louder and louder until he
is about two meters to the right of dead center."[17]

The sketches in Figure 5.4 show what is happening (to make
it easier to see, this drawing uses a circle rather than a complete

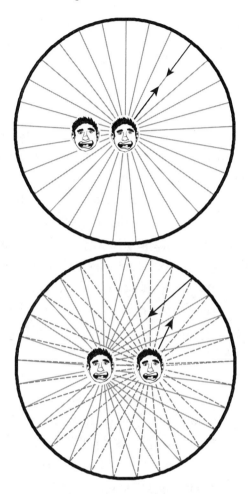

Figure 5.4 Focusing in the Mapparium.

sphere). When the talker speaks from the center (top diagram), all the reflections are focused back, so the talker appears to be surprisingly quiet to the listener left of center. If the talker moves to the right, then the place where the reflections focus moves closer to the listener. The sound will be loudest when the talker and listener are arranged symmetrically about the center (bottom diagram).

Although this effect is particularly strong in the Mapparium, which has curved surfaces above, below, and in front of the speaker, given the right structures it can even be heard outdoors, as reported by José Sánchez-Dehesa from the Polytechnic University of Valencia in Spain.[18] The Cempoala (Zempoala) archaeological site near Veracruz in Mexico is one of the most complete surviving examples of an Aztec ceremonial center. It is a large grass plaza dotted with the remains of various structures, including a low, circular enclosure topped with merlons. The guidebook hedges its bets about the purpose of the structure: "associated with gladiator worship of Mexica[n] (Aztec) origin, although it may have served as an intake for rainwater." In pictures, the structure looks like a stone sheep enclosure made of large, round pebbles. Whatever the purpose of the circle, it displays an acoustic focus. Sánchez-Dehesa reports that if a listener stands in the right place while a companion walks and talks along a diagonal through the circle, the sound gets louder as the speaker moves farther away.

Hoping to visit a spherical room, I sought out urban explorers, people who like to illicitly explore sewers, abandoned subway stations, and derelict buildings, enjoying the visceral thrill of trespassing in ghostly places where others do not go, and finding hidden histories and traces of previous occupants.[19] One member of Subterranea Britannica, a group dedicated to the legitimate exploration of underground spaces in the UK, sent me an e-mail describing a dome in Berlin that was one of the most important listening stations for the West during the Cold War. The enclosed photo showed the dome on top of a derelict tower; this was a place I had to visit.

The abandoned spy station sits atop Teufelsberg ("Devil's Mountain"; Figure 5.5), which rises up from the Grünewald forest. As I walked up through the woods on a hot summer's day toward the listening station, it seemed unbelievable that this large hill was man-made. It was constructed from millions of cubic yards of rubble created by bombing raids and artillery bombardments during World War II.[20]

Figure 5.5 Teufelsberg.

Before entering, I had to sign a liability waver because the derelict buildings have many holes and missing walls with no protection from the shear drops below. My German tour guide was Martin Schaffert, a young history scholar sporting a neat beard, small ponytail, glasses, and a flat cap. As Martin explained the history of the site, I surveyed what remained of the buildings. Doors and walls were missing, with the debris of the crumbling buildings lying on the floor mixed with broken glass from semi-illicit parties. Where walls were intact, they were covered in graffiti. My eye was drawn upward, where on top of the main building were three domes; some were vandalized, with their walls partly in tatters, but the topmost one, sitting on top of a five-story tower rising out of the roof, was intact.

These were *radomes* (a contraction of the words *radar* and *domes*), used to hide the spying activities from prying eyes as the British and Americans listened in on broadcasts and wireless communications from East Germany, Czechoslovakia, and the Soviet Union. The spherical domes were also used to protect the listening equipment from the elements, especially wind and ice. All that remains now are the concrete plinths to which the antennae were bolted. The domes were constructed of triangular and hexagonal fiberglass panels stretched between a scaffold, looking like giant soccer balls. Fiberglass is transparent to electromagnetic waves and so ideal for radomes; this was one of the reasons the material was developed during World War II.

The tower supporting the top dome was missing all its walls, but the dome itself was virtually complete because it had been rebuilt and reused for air traffic control over Berlin. Every surface of the dingy stairwell up the center of the tower was coated in graffiti. As I climbed the stairs to the dome, I could hear the reverberated voices of other visitors enjoying the acoustics. With a midfrequency reverberation time of about 8 seconds, the radome sounds a bit like a cathedral. Musicians come to perform music in the space. But there is more to enjoy than just the reverberance.

Once inside, I stood and watched other visitors; it was wonderful to see their faces light up when they realized how odd the acoustic was. The slightest sound, even a simple footstep, created a ricocheting effect. Some were prompted to explore with gusto (a good hard stomp could be heard to repeat eight times, sounding like distant firecrackers), while most contented themselves to playing more subtly, almost treating the space with the same reverence as they would a place of worship.

I then climbed onto the old antenna plinth to get to the center of the room. The dome is roughly the top two-thirds of a sphere about 15 meters (50 feet) in diameter, constructed from yellowing hexagons. A band of graffiti 2 meters (about 7 feet) high ran above the floor, interrupted only by a second small opening, through which was an unprotected shear drop to the roof of the building five stories below. I got out my recorder to dictate my impressions of the place, and noticed how every word was doubled up by the reflections from the radome.

Here in Teufelsberg I wanted to explore an effect that happens in the spherical Mapparium.[21] The unusually strong focus makes it possible to experience the strange sensation of whispering into your own ear. Or, as Hartmann puts it:

> As you approach the exact center of the Mapparium sphere you suddenly become aware of strong reflections of your own voice … If you sway to the left, you hear yourself in your right ear. If you sway to the right, you hear yourself in your left ear.[22]

In Teufelsberg the effect is strongest if you look upward while whispering, because there is a larger concave area to focus the sound. Thus, with my head tilted backward, five stories up in a fiberglass radome in Berlin, I discovered an exciting binaural sonic wonder, an effect that reveals how we work out where sound is coming from. Having two ears enables mammals to perceive the locations of sound sources. Hearing evolved to enable animals to sense danger, alerting them to predators sneaking up and trying to turn them into lunch. Humans have good vision, but our eyesight cannot detect threats from behind, so being able to hear and locate danger is crucial.

There are a couple of main ways in which we sense where sound is coming from. Imagine that someone is talking to you from your left side. The sound arrives first at your left ear because it takes slightly longer to reach your right ear. Your brain also makes fine distinctions in loudness. The sound has to bend around your head to reach the right ear, causing it to be much quieter at high frequency. (The loudness of low frequencies coming from far away is minimally affected by your head itself). Your brain compares the timing in each ear at low frequency and the relative level of high frequencies to decide where the sound is coming from.

Spherical rooms can mess with both of these cues. The loudness cue can be distorted, creating an unexpected localization. It makes sound appear to come from the wrong direction, as Hartmann describes: "Suppose you are on the Mapparium bridge facing South America. There is a source of noise to your right, but you discover that you hear the noise coming from your left!"[23] Strong, focused reflections from the sphere create a loud reflection at the left ear, fooling the brain into localizing the noise on the left side.

Localization is usually based on the first sound arriving at the ears (the precedence effect). This rule of thumb serves us well because the earliest sound takes the quickest route, which is usually a straight line between talker and listener. You may have sat through a church service in which the sermon appeared to be delivered by the loudspeakers rather than by the preacher. The reason for this impression is that the sound from the loudspeakers reaches the listener first. Adding a little delay electronically in the public address system, so that the first sound wave to reach the listener comes directly from the preacher's lips, solves this problem.

But adding a delay is ineffective if the loudspeakers are turned up too high, because the precedence effect can be overruled by a loud sound that arrives later—a situation that prevails at most rock concerts. Without electronic amplification, though, reflections from walls are usually too quiet to cause a problem. But in the case of the Mapparium or Teufelsberg, where the dome's focus greatly amplifies the delayed sound, the reflections are so strong that we are fooled by false localization. When I burst my balloons in Teufelsberg, the first reflection from the ceiling was 11 decibels louder than the sound coming directly from the balloon (Figure 5.6). (A useful guide is that an increase in 10 decibels is roughly a doubling in

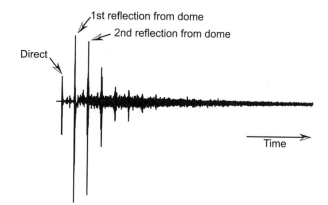

Figure 5.6 Direct sound and reflections from bursting a balloon in the center of the Teufelsberg radome.

perceived loudness.) When I knelt down to unzip my knapsack, it sounded like someone was opening the bag above my head!

Barry Marshall, from the New England Institute of Art in Brookline, Massachusetts, who used to be a guide at the Mapparium, told me how he used the acoustic to play practical jokes on visitors and "blow their minds." The strong focus meant that he could stand far away from visitors and surprise them by calling "over here," and they would look in the wrong direction.[24] In Teufelsberg, I contented myself with spying on other conversations by finding the point where the speech of other visitors was being focused.

Long-distance whispering and sound focusing tend to unnerve people because they hear something that seems supernatural. If I were to chat with you in a normal room, the low frequencies in my voice would have roughly the same loudness in both your ears, whichever way you faced, because they can easily bend around your head by diffraction. Normally, low frequencies become much louder in just one of your ears only if I get very close, maximizing the "sound shadow" cast by the head. The effect reduces low frequencies in the far ear, making you think I am next to you. But in the Mapparium, the sphere can focus sound so intensely on one ear that it can trick the brain into thinking that I must be close by. Not only can I whisper sweet nothings to a loved one yards away; I can even narcissistically whisper them to myself!

The point of whispering, of course, is to say something so quietly that unintended listeners cannot hear it. Apparently, the original dome of the Capitol allowed members of the House of Representatives to whisper private messages to each other. But the amplification worked both ways: the congressmen could also overhear their colleagues' secrets. We naturally seem to associate sound amplification by curves with spying, subterfuge, or illicit liaisons. Fellini used this link to dramatic effect in his film *La Dolce Vita*, in which a concave basin allows eavesdropping on conversations from the lower floor of a villa.[25] But the most curious spying legend concerns a large limestone cave near Syracuse in Sicily called the Ear of Dionysius. The story goes that the tyrant Dionysius (ca. 430–367 BC) used the cave as a prison and exploited the acoustics to find out what hapless detainees were whispering to each other.

The cavern is shaped like a tall, pointy donkey's ear and narrows dramatically at the top. The wedge shape acts like a funnel to sound, as Figure 5.7 illustrates, potentially collecting whispers from the ground level and concentrating them at the roof of the cave, 22 meters (72 feet) above. Legend has it that Dionysius spied on prisoners from a listening chamber at the top, picking up the amplified sound through a small hidden opening at the top of the cavern.

The cave is a popular tourist venue, and in the past, the listening chamber could be visited—as one traveler noted in 1842, "the only . . . mode of access being by means of a rope and pulley, the adventurer hazarding his life in a little crazy chair."[26] Despite the legend being recounted to tourists, some reports cast

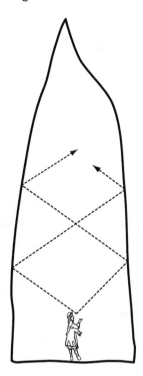

Figure 5.7 Sound in the Ear of Dionysius.

doubt on whether spying was possible. In 1820 the Reverend Thomas Hughes wrote, "A very low whisper is heard only as an indistinct murmur; the full voice is drowned in the confusion of the echoes. The voices of several persons speaking at the same time are as unintelligible as the cackling of geese, so that if the ancient Sicilians were half as loquacious as the modern, who always chatter in concert, they must very often have put the listening tyrant to a nonplus."[27]

With modern health-and-safety precautions, visitors can no longer enter the upper chamber. A listener today is simply left to enjoy the reverberation at ground level, marvel at the legend, and gaze at the cavern's large, ear-like shape. (A different sound-related tradition that has been stopped is the firing of firearms for tourists: according to another nineteenth-century visitor, "A pistol was fired, and the report was like the discharge of an eight-and-forty pounder."[28])

Recently, Gino Iannace, from the Second University of Naples, and collaborators persuaded the cave owners to let them into the listening chamber to survey the acoustic. Just as my team rates theaters, classrooms, and railway stations, Iannace's group took a battery of measurements to assess speech intelligibility in the cave. The results were "on the bad side of average," indicating that the cave's reverberation makes speech muddy and incomprehensible. Undeterred, Iannace then carried out a series of perceptual tests, asking listeners to transcribe phrases recounted in the cave, but nobody could get a single word right. Disappointingly, the scientific measurements failed to support the legend.

———

Utterly overcome by pain and grief, I crouched against the granite wall.

I just commenced to feel the fainting coming on again, and the sensation that this was the last struggle before complete annihilation—when, on a sudden, a violent uproar reached my ears. It had some resemblance to the prolonged rumbling voice of thunder, and I clearly distinguished sonorous voices, lost one after the other, in the distant depths of the gulf.

Suddenly my ear, which leaned accidentally against the wall, appeared to catch, as it were, the faintest echo of a sound. I thought that I heard vague, incoherent and distant voices. I quivered all over with excitement and hope!

"It must be hallucination," I cried. "It cannot be! it is not true!"

But no! By listening more attentively, I really did convince myself that what I heard was truly the sound of human voices.[29]

This is the moment at which Professor Hardwigg and Harry, the heroes of Jules Verne's *A Journey to the Centre of the Earth*, miraculously reconnect via a whispering wall formed from a granite labyrinth. It is a prodigious structure, and Harry works out that he is hearing Hardwigg 8 kilometers (5 miles) away.

Outside Jules Verne's imagination and aboveground, the largest real whispering wall I know of is 140 meters (460 feet) long—a mere baby in comparison, and also less poetic; it is the concrete dam that withstands the Barossa Reservoir in South Australia. For some reason the dam was built to be a precise arc. This vast, gray slab of concrete has turned into an unlikely tourist attraction, with visitors chatting with each other from opposite ends of the dam.

This wall does not focus sound as happens with elliptical ceilings and domes; the listener and speaker are too far from the focal point of the arc. What happens is that the sound hugs the inside of the concrete wall and is transported with surprising loudness to the other side of the dam.[30]

Whispering arches behave in a similar way, and they also show up in the most unlikely of places. On the lower level of Grand Central Terminal in New York City, outside the famous Oyster Bar & Restaurant, sweeping tiled archways, designed by Rafael Gustavino and his son in 1913, support the ceiling. If you whisper into one side of the arch, the sound follows the curve of the tiled ceiling before coming back down the other side. For the best effect, the whisperer and listener need to get close to the stone, like naughty children standing in opposite corners of a classroom.

This scene does not immediately make me think of marriage proposals, but still the location is a popular place for popping the question (the jazz musician Charles Mingus was supposed to have done so there). The sound effect has also inspired literature and films; the author Katherine Marsh uses the whispering arches as the starting point in her children's books *The Night Tourist* and *The*

Twilight Prisoner, describing the arches as "one of the coolest places in New York."

I have found a dozen documented cases of whispering arches. Very few, if any, seem to have been designed for their sonic quirkiness. The arch in St. Louis Union Station, Missouri, is adorned with a plaque that begins: "The Whispering Arch, an architectural accident or the sharer of secrets?" (a curious question, since presumably it could be both). Apparently, the sound effect was discovered in the 1890s, when, as the plaque says, "a workman dropped a hammer on one side of the arch and a painter on the other side, nearly 40 feet [12 meters] away, heard him." This whispering arch was thus an accident of geometry.

No doubt there are many other whispering arches to be found. Elaborate architraves around doorways that contain sweeping arcs help channel sound from one side to another. Acoustician and Mayan pyramid expert David Lubman measured one in West Chester University in Pennsylvania. The doorway had an arc rather like a piece of upside-down, bent guttering within which sound propagates. People are so used to hearing sound become quieter as they get farther away that they find whispers emerging from the half pipe of the arch surprisingly loud. Lubman wonders whether this whispering feature was deliberate, because the half pipe that carries the sound seems to have little other purpose.[31] But it may just have arisen by accident—a by-product of the door design. Sadly, the feature is largely ruined by traffic noise.

My favorite whispering arch is at the ancient monastic site of Clonmacnoise in County Offaly, Ireland. (How can a collector of sonic wonders resist that name?) An ornate Gothic doorway dating to the fifteenth century has carvings of Saints Francis, Patrick, and Dominic above and opens into the roofless remnants of the cathedral. Like the Oyster Bar archway in Grand Central Terminal, it is a popular spot for wedding proposals. Folklore has it that the door-

Figure 5.8 The architrave half pipe you whisper into at Clonmacnoise.

way once had a very unusual use: Lepers would stand at one side of the doorway and whisper their sins into the half pipe in the architrave (Figure 5.8). The priest would stand at the other side of the arch, far enough away to avoid infection, listening to the confession emerging from the architrave. I spent an afternoon watching busloads of foreigners having fun whispering in the arch, despite rain and howling winds. How do these whispering arches work? They behave like whispering galleries.

The whispering gallery of St. Paul's Cathedral in London gave me, as a young teenager on a scout trip, one of my first acoustic memories. The cathedral is built in the shape of a cross, with the dome rising above the intersection between the arms. It is such an important landmark of London that, during the Blitz in World War II, Prime Minister Winston Churchill ordered it to be protected at all costs, to boost morale.

Visitors climb 259 steps from the main cathedral floor up to the base of the dome and then emerge onto a narrow floor only a couple of yards wide running around the inside of the dome's circular walls. At this point the dome has a diameter of 33 meters (108 feet). Metal railings line the inside edge of the gallery to keep people from falling off as they look up to the top of the dome or down to the cathedral's floor while admiring the opulent splendor. I remember having great fun calling out to friends around the dome. It was busy and noisy, yet I could still hear my friends' rude whispers carry a remarkable distance.

Whispering galleries have fascinated many famous scientists, such as Astronomer Royal George Airy, best known for his work on planetary science and optics. In 1871 he published a theory of how whispering galleries work, but it explains what happens only in perfectly spherical rooms such as the Mapparium. The Nobel Prize–winning physicist Lord Rayleigh was also intrigued, writing that "Airy's explanation is not the true one" for St. Paul's. To prove his point, Rayleigh made a scale model of a whispering gallery from a semicircular strip of zinc 3.6 meters (12 feet) long.[32] He used a birdcall whistle at one end to produce a chirp skimming along the inside of the metal strip; at the other end the sound was remarkably strong, powerful enough to make a flame flicker. But when a narrow barrier was placed anywhere along the inside wall of the zinc strip, the flame remained undisturbed. This result showed that the sound waves were hugging the inside surface of the curved strip.

Sound clinging to and following the inside of the gallery walls is a pleasing scientific finding, but that alone does not explain the startling effect of the whispering gallery. Visitors often hear peculiar sounds, as reported by C. V. Raman in his 1922 paper:

> In response to ordinary conversation, strange weird sounds and mocking whispers emanate from the wall around. Loud

laughter is answered by a score of friends safely ensconced behind the plaster. The slightest whisper is heard from side to side, and a conversation may be easily carried on across the diameter of the dome, in the lowest undertone, by simply talking to the wall, out of which the answering voice appears to come.[33]

The sound hugging the wall creates an aural illusion because it is much louder than expected. In addition, both whisperer and listener need to get close to the wall. When listeners move their ear a short distance away from the wall, the sound suddenly gets much quieter. When the brain tries to work out how far away a sound source is, it uses loudness as a clue. Normally whispers are loud only when you are close to the talker. Furthermore, only when a source is close by does the loudness decrease rapidly with small movements of the head. The brain misinterprets the rapid quieting of the whispers as the ear moves away from the wall, and it thinks the source must be in the stonework.

Raman received a Nobel Prize for his work on light scattering, but he also carried out extensive research in acoustics. In the early twentieth century he documented five different whispering galleries in India, including the vast seventeenth-century Gol Gumbaz mausoleum in Bijapur. From the outside, Gol Gumbaz is an imposing building, testifying to the power of the Adil Shahi Dynasty; it rises impressively from the surrounding plain. It looks like a giant cube, with slender octagonal turrets in the corners and a vast dome nearly 38 meters (125 feet) in diameter on top, some 30 meters (100 feet) from the floor. In the words of the acoustic engineer Arjen van der Schoot:

If you walk into the place the internal size of it is humbling, but you soon forget because you'll be taken by the acoustics as soon as you start making sound. The reverb[eration] in Gol Gumbaz is so staggering that Indian people travel for days just to hear it. And when they arrive they find a hundred people inside screaming at the top of their lungs.[34]

With children enjoying yelling and listening as their voices repeat over and over again, the atmosphere is like a crowded day at a swimming pool. Van der Schoot was carrying out acoustic measurements and so had the rare pleasure of enjoying the mausoleum when empty: "It took us two years to get the proper permissions and clearing the place for a couple of hours. Bus loads of people were held at the gate so we could do our work in silence in the astonishing whispering galler[y], where, when it is silent, you can count 10 echoes to your whisper."[35]

As much pleasure as Gol Gumbaz gives its visitors, the whispering gallery was an accident of design. The decision to put a dome on top of the hall was made only after building had begun. I can find evidence of only one whispering gallery constructed deliberately. According to a 1924 edition of *Through the Ages Magazine*, "The whispering gallery in the Missouri State Capitol [Jefferson City, 1917] was carefully, mathematically laid out by a celebrated expert on acoustics, and this is undoubtedly the first time on record that such a thing has been done successfully."[36]

For an acoustics conference, I produced some animations of how sound waves move around a whispering gallery. Using the latest algorithms on a fast computer, the movie showed how whispers are carried around by hugging the walls. In an idle moment while preparing my talk, I dashed over to the library to borrow a copy of the nineteenth-century acoustic bible *The Theory of Sound*, by Lord Rayleigh, which, remarkably, he wrote while recovering from rheumatic fever in Egypt. His description of how whispering galleries work is simple to sketch out—much easier than using my complex computer models.

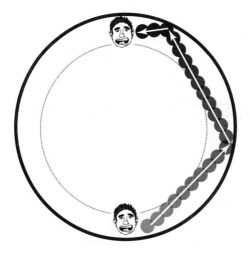

Figure 5.9   Sound in a whispering gallery.

Imagine cuing a ball on a circular pool table so that it goes off almost parallel to the side wall. The ball then maps out how sound moves around the gallery when someone whispers along the wall. An unexpected effect becomes apparent: the ball hugs the wall, circulating close to the curve without ever going into the middle of the circle. The same thing happens with sound in a whispering gallery, as shown in Figure 5.9.

When I visited the listening station at Teufelsberg, I demonstrated the whispering-gallery effect to my tour guide, Martin. He had encouraged others to try out the acoustics in the center of the room but was unaware that the voices could skim around the edges. Later, in brief moments when the radome was empty, I made a measurement by bursting a balloon on one side of the dome, while my sound recorder was balanced next to the wall on the other. With such a loud bang, the sound can make many circuits around the edge of the dome before dying away to silence. For one burst I counted eight clear bangs. A plot from one of the recordings (Figure 5.10) shows four or five spikes at times when the bang passes the microphone.

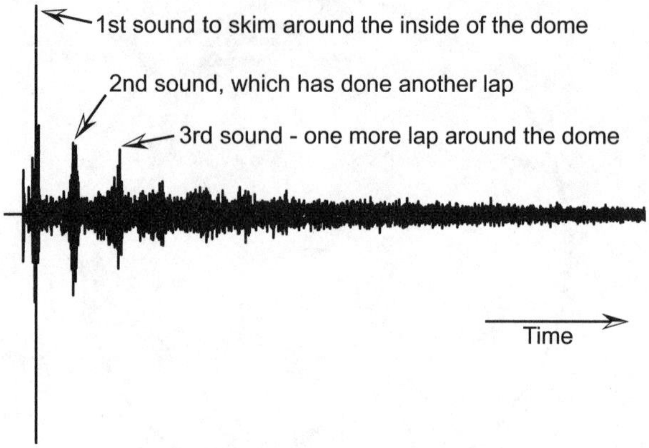

1st sound to skim around the inside of the dome

2nd sound, which has done another lap

3rd sound - one more lap around the dome

Time

Figure 5.10 Sound created by bursting a balloon in the Teufelsberg radome being used as a whispering gallery.

But why is St. Paul's a "whispering" gallery and not a "speaking" gallery? I recently returned to the cathedral and made some covert recordings. The gallery is best visited early in the day, when there are not too many other people making noise. It is also best visited with a friend who can do the whispering, but I was on my own. Fortunately, the attendant was particularly adept at producing just the right whispers. Back in my laboratory I analyzed the recordings, which showed one good reason for whispering rather than talking normally: The background noise leaking up from the main floor of the cathedral is quite loud across the bandwidth of normal speech. But in the higher frequencies at which the attendant was whispering, the background noise is much quieter, so the ghostly sounds are not swamped by the background hubbub.

Most of the large sonic wonders that I have discovered were accidental, but what sounds could be made if we really tried? What shapes could exploit the physics found in accidental sonic wonders to create new aural effects? One could take inspiration from the seventeenth-century Jesuit scholar Athanasius Kircher. In addition to writing about the infamous cat piano, he imagined and drew fantastical acoustic contraptions such as speaking statues and a music-making ark to mechanize music composition. Perhaps modern inventors should come up with contemporary versions of these sketches.

While traveling the world looking for sonic wonders, I have begun dreaming up a few of my own. Investigating the distortions from the spherical radome in Teufelsberg made me think of a fairground years ago when I played with fun-house mirrors. One was curved in a way that turned me into a distorted goblin. Another was bent differently, creating a disfiguring reflection that stretched out my legs so that my torso almost disappeared. Could I design a whispering gallery using a complex curve? Such a design would be new; the whispering galleries and walls I have discovered have all been simple arcs, curves, or domes.

In his paper on whispering galleries, physicist C. V. Raman describes Golghar, the old government granary in Bankipore, India. Built in 1783, it is shaped like a beehive, and at 30 meters (100 feet) high there are great views to be seen from the top. Raman notes of the sound inside, "As a whispering gallery there is perhaps no such building in the world. The faintest whisper at one end is heard most distinctly at the other."[37] But it was photos of the outside of the building that intrigued me. An external spiral staircase hugs the outside like an old-fashioned amusement-park slide. If the outside wall of the staircase were altered to gently curve, sound could spiral up the walls of the staircase. It would be a whispering "waterslide," to complement the whispering gallery inside.

The computer program that produced animations of St. Paul's Cathedral enables me to design odd-shaped whispering galleries and test whether they will work. This is how acoustic engineering is done nowadays. Before places are built, computers are used to

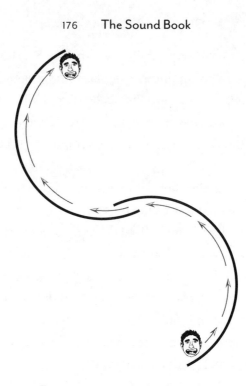

Figure 5.11 New whispering walls.

check whether an actor will be heard clearly by the audience, or whether a public address system in a railway station will produce intelligible announcements. Thus I turned my engineering skills and scientific knowledge on their head. Instead of designing to remove acoustic aberrations caused by curved surfaces, I used the same tools to maximize aural distortions.

Some of Richard Serra's artworks in the Guggenheim Museum in Bilbao, Spain, are giant steel walls that behave like whispering walls. Taking inspiration from these works, I thought about the potential design. I wanted whispers to hug an S-shaped curve like a fun-house mirror, but unfortunately, sound does not hug a convex section. I solved this problem by forming the S shape from two arcs (Figure 5.11). With that arrangement the sound passes around the inside of the first curve and jumps the small gap separating the two sheets before skimming around the inside of the second bend to be heard by the listener as surprisingly loud.

The delight in these places comes from hearing a voice carry an unexpected distance, and this effect is more dramatic if the sound is a quiet whisper to begin with. Later mathematical analysis by Lord Rayleigh suggests another reason for whispering: high frequencies, like the sibilant tones in whispers, hug the walls closer than the lower-frequency sounds of normal speech. It also seems that the St. Paul's gallery is particularly adept at transporting sound. Acousticians have attributed this feature to the slight sloping of the walls. By tilting the walls inward at the top, less sound goes upward and is lost to the top of the dome.

I now had the answer for why the sewer had made my voice spiral. While the analysis by Lord Rayleigh shows that the hugging effect in whispering galleries is more pronounced for larger circles, the theory also demonstrates that the same circulating effect happens in smaller places, even tunnels only a couple of yards across. With my head close to the ceiling of the sewer, my speech simultaneously disappeared down the tunnel and skimmed and circulated around the curved walls as happens in a whispering gallery. I was not hearing an audio illusion in the sewer; the sound was genuinely spiraling.

6

Singing Sands

Ayear after my visit to St. Paul's Cathedral, I traveled to Kelso Dunes (Figure 6.1) in the Mojave Desert, California, with sound recordist Diane Hope in the hope of hearing a sand dune sing. Kelso is one of about forty documented sites where this happens.[1] The English naturalist Charles Darwin recounts tales of the "El Bramador" hill in Chile, which was called the "roarer" or "bellower" by locals.[2] Ancient Chinese writings describe festivities at the Mingsha Sand Dunes: "It is customary on the tuan-wu day (the Dragon festival on the fifth of the fifth moon) for men and women ... [to] rush down ... in a body, which causes the sand to give forth a loud rumbling sound like thunder."[3]

A sand avalanche starts the dune singing. The slope has to be steep, and the sand has to be very dry. But dry sand is by definition the loosest, which meant my feet struggled to find purchase on Kelso's surface. I had prepared for the extreme temperatures of the summer desert, but I had not realized that searching for musical dunes was going to be an aerobic workout. I wanted to gasp for air as I struggled up the dune, yet I had to hold my breath so as not to ruin the audio recordings.

As I trudged up the sandy slope, my feet made a burping noise in the sand. It reminded me of the first part of Marco Polo's description. He wrote that dunes "at times fill the air with the sounds of all kinds of musical instruments, and also of drums and the clash of arms."[4] I was getting nothing dramatic like drums, but I was creating something musical. Each labored footstep created a single honk, like a badly played tuba. Toward the top of the slope I got so tired that I resorted to scrambling up on all fours, producing a comical brass quartet.

While the burping was entertaining, I was frustrated because

Figure 6.1 Kelso Dunes.

the dunes were not in full voice. What I had traveled to hear was a sustained boom that supposedly reaches 110 decibels, resembling the loudness of a rock group, audible up to a mile away.[5] It was getting toward late morning. The wind was making recording difficult and the heat was becoming unbearable, so we retreated down the dune to try again the next day.

Back at the campsite, I listened to a recording of my phone interview with Nathalie Vriend from Cambridge University, who had studied the singing sands for her doctorate. I was looking for hints of how to find the best place on the dunes. Worryingly, in the middle Nathalie mentioned that a friend had visited Kelso in recent times and had been disappointed by the sound. I also reviewed key scientific papers. I hoped that, by understanding the physics, I might have a better chance of getting the dunes to properly sing the next day. While scientists agree that burping sand is a necessary ingredient, there is a heated debate about what actually causes the loud booming. Is a deep layer of the dune vibrating like a giant musical instrument? Or are the sand grains locked in a synchronized avalanche?

During a proper boom, thousands of sand grains sing in a coordinated choir stretched over many yards across the dune. Waterfalls are similarly made from an extended orchestra, but the instrumentalists are tiny bubbles. The loudest waterfall I have ever heard is Dettifoss on the glacial river Jökulsá á Fjöllum in Iceland; it is also the most powerful in Europe. Many years ago, my wife and I cycled there on a bitterly cold and unpleasant morning. The road was more like a track, rough and potholed, and the northerly headwind poured cold air down from the Arctic with such strength that it brought us to a standstill at times.

We made tortuous progress, first through winding moorland and then through the sandur—a desolate landscape of glacial outwash and black volcanic silt where little grows. Leaving the bikes behind, we gingerly walked to the cliff edge overlooking the falls, which are more than 100 meters (325 feet) wide and 44 meters (145 feet) high.[6] Standing very close to the precipice, a pang of fear swept through my body: with 180 cubic meters (6,000 cubic feet) of water pounding over the edge per second, a slip now would end in certain death. The incessant, thunderous power meant we had to shout to be heard. The noise seemed to cover every frequency at once, from a bass rumble up to a high-pitched hiss. The falls presented an overwhelming and isolating noise, like the sort controversially used by the CIA to achieve sensory deprivation during interrogations.[7]

Water might be a simple substance, but it can make a vast range of sounds, from babbling brooks to crashing waves, from torrential rain to the plink of a single drip. Describing Yosemite Falls, the American naturalist John Muir wrote that the water "seem[s] to burst forth in irregular spurts from some grand, throbbing mountain heart... At the bottom of the fall... [i]t is mostly a hissing, clashing, seething, upwhirling mass... This noble fall has far the richest, as well as the most powerful, voice of all the falls of the Valley, its tones varying from the sharp hiss and rustle of the wind in the glossy leaves of the live-oak and the soft, sifting, hushing tones of the pines, to the loudest rush and roar of storm winds and thunder among the crags of the summit peaks."[8]

For decades, scientists have been interested in how falling water, such as crashing waves, creates sound underwater, because the noise impedes submariners listening for the enemy. But I want to know about what happens above the surface, and fortunately, scientists have now turned their attention to this question.

Laurent Galbrun from Heriot-Watt University in Scotland has been investigating how to make an impressive-sounding fountain or water feature while pumping the minimum amount of water, to reduce energy use. In parallel work, Greg Watts and colleagues at

Bradford University in England have been investigating water falling onto different rocks and into pools, in search of the best sounds for hiding traffic noise. After recording a diverse playlist of water features, they had a panel of listeners judge the pleasantness of each sound. This experiment had to be undertaken in an acoustic laboratory, a room hardly conducive to making aesthetic judgments on outdoor water features. So the researchers built a theater set within the laboratory. It was a garden balcony, complete with bamboo screens, potted plants, and garden furniture, to get the subjects into the right frame of mind.

After the subjects had scored how much they liked each sound, Watts came to the conclusion that the worst noises had a booming quality, reminiscent of water flowing down into drains or utilitarian culverts. The most pleasing sounds sploshed and splashed, having a natural randomness as the water fell onto an uneven surface formed from small boulders. In similar tests, Galbrun found that the gentle babbling of a slow-moving natural stream was the most relaxing of all water sounds tested.[9]

The source of the waterfall sounds surprised me at first. A television crew recently filmed what happens in my university's anechoic chamber. A high-speed camera captured a single drop falling into a fish tank of water. A slow-motion video from the top looks pretty. The drop causes a narrow column of water to rise up from the surface, creating ripples. But to understand what is heard, you need to look from the side, just underneath the surface of the water. While the ripples are visually impressive, a single tiny air bubble generates most of the sound. As the bottom of the drop penetrates the surface of the water, a bulging meniscus is created from which a tiny air bubble suddenly breaks away. This bubble of trapped air is only a few millimeters in diameter and so is easy to overlook and difficult to film. It may be small, but the air inside the bubble vibrates, resonates, and creates a plink that travels through the water and into the air.

Water falling onto rocks sounds very different because no underwater bubbles can be made (unless a layer of water has built up on the stone). Again, it is easiest to think about what happens when one drop smashes onto a rock and is splattered across the stone. As the falling drop smears itself into a thin layer of water on the stone, it disturbs the air around it, creating the sound.

A couple of months after the television crew filmed a single bubble, I got to learn more about aquatic sounds from artist Lee Patterson. We met up in the English Lake District, where Lee described how, in local ponds and water courses in northern England, he had discovered underwater sounds as rich as tropical rainforests. We chatted about the piece he was going to compose from his recordings in the Lake District. *The Laughing Water Dashes Through* was going to be a work prompted by the devastating floods that had hit the nearby market town of Cockermouth a few years previously. Lee explained how the work would explore the "different forms of energy embodied by water flow, and the sound that happens as a by-product of the water flow."[10]

He was recording in a small, enclosed, flooded quarry on the day I visited. With blazingly hot sunshine and the birds singing around us, it was an idyllic spot (provided we stood with our backs to an ugly concrete shed). Lee had simple homemade hydrophones, constructed from a sliver of shiny piezoelectric material, which makes electricity when it is deformed by underwater sound waves, embedded in the tops of brightly colored plastic screw caps from pop bottles. He cast these into the water, turned up the amplifier, and passed me the headphones.

I heard a malevolent munching and crunching. It was as if an animal was trying to nibble away at my eardrum. The sound came from tadpoles scraping the hydrophones, in the vain hope that there was algae on the bottle caps. The tadpoles were swimming among oxygenating pondweed, and with a careful repositioning of the hydrophones, strange mechanical chirps could be heard, like bacon being deep-fried. These were caused by a fast stream of small bubbles rising up from the pondweed, looking like champagne bubbles rising up in a glass. It turns out that photosynthesizing plants were making the bubble streams.[11]

A few days later, I spoke to Helen Czerski from the University of Southampton, who studies how sound is made as bubbles are created. Her research shows that as bubbles form at a small nozzle, sound is created because the bubble initially has a teardrop shape while attached to the nozzle, but when it breaks away into the body of the water it forms a sphere. This shape change causes the bubble to vibrate, resonating the air inside and creating sound. Helen was skeptical that this is what happens with pondweed because natural bubbles from photosynthesis form more slowly and therefore probably lack the impulsive kick as they break off. She thought it was more likely that I was hearing the bubbles bumping into each other or into the hydrophones.

The waterfall at Dettifoss in Iceland can be explained by scaling up the effects of a single oscillating bubble, to consider the vast number of bubbles in the white cascade. Each ball of trapped air is a different size and plinks a note at its own particular frequency. In the cascade, the combination of millions of random plinks creates a vast bubble orchestra, which fizzes and roars.

Each waterfall has its own voice. If it tends to have lots of larger bubbles, it will have a bassy rumble. Smaller bubbles result in more hissing, like the Yosemite Falls described by Muir. Surrounding rocks can further alter the sound. Svartifoss in southern Iceland is only about 20 meters (65 feet) high. It has water cascading from a horseshoe of overhanging cliffs made up of hexagonal basalt columns. The name, meaning "black fall," comes from the color of the rocks, and on the day I visited, the color was strongly emphasized by the overcast, drizzly weather. However, it is worth an hour's walk even in the rain, because not only do the surrounding rocks make stunning holiday snapshots, but they also amplify the water as it slaps and hisses against them.

Another impressive Icelandic waterfall is Seljalandsfoss, where you can go behind the curtain and be surrounded by noise, as the fizzing from the water hitting the pool is reflected from the cliff behind you. The flow of water is not constant, which means the noise splutters. Close your eyes, and you can imagine a small freight train rumbling past overhead.

While waterfalls are common, the sound of a tidal bore, a single tall wave that sweeps inland up a narrowing estuary, is much rarer. The bore of Rio Araguari in Brazil is named *pororoca* in the language of the native Tupi, which translates as "mighty noise."[12] Closer to my home is the Severn bore, near Gloucester in England. Early on a misty September morning, part of a brief Indian summer, forecasters predicted a four-star bore on the heels of a large tide driven by the autumnal equinox. As I wandered along the banks of the river, I saw a few surfers midstream, clutching their boards ready to catch the wave; this must be a good place to watch, I thought. I first stood close to the water's edge but then realized that the silt all around me was from the previous night's tide, so I retreated higher up the bank. You have to be careful when dealing with tidal forces. In China, eighty-six people were swept away by a tidal bore on October 3, 1993.[13]

Then I waited, waited some more, and then waited even longer. Twenty minutes behind schedule, a rumble started downstream. The bore came into view and broke on the opposite bank, forming a continuously breaking wave right across the river. It resembled a big ocean breaker, but instead of the soothing rhythm of waves crashing on the shore one after another, the bore produced the continuous sound of a breaking wave.

Right behind the Bay of Fundy in Nova Scotia, the Severn Estuary has the second-highest tidal range in the world—as much as 14 meters (45 feet) for spring tides. A map of the River Severn shows its sinuous funnel shape. What the map does not show is that the depth of the river decreases rapidly as you go inland. When the huge tide enters the estuary at the sea, the water is forced up the narrowing channel, which gets shallower and shallower. The excess water can go only one way, upward, thus forming the surge wave.

While the first wave is the star of the show, if you rush off too soon you miss the sound of the "whelps," the secondary undulations that follow the bore. Floodwater surges behind the bore for a good thirty minutes after the main wave, the force of which is apparent as it pulls along whole trees and other debris. Large undulations form in the water. These waves break here and there, creating a crashing sound to accompany the gurgling and rumbling of the huge mass of water being moved—an audio mixture of waves on a beach and water running down a municipal drain.

In terms of bore heights, the River Severn comes in fifth, with larger ones, like the *pororoca* in Brazil, having an even more dramatic sound. The Qiantang River bore was described by the Chinese poet Yuan as, "10,000 horses break out of an encirclement, crushing the heavenly drum, while 56 huge legendary turtles turn over, collapsing a snow mountain."[14] In 1888, W. Usborne Moore, a commander in the Royal Navy, described it in a more understated way: "On a calm, still night it can be distinctly heard, when 14 or 15 miles distant, an hour and twenty minutes before arriving. The noise increases very gradually, until it passes the observer on the bank of the river with a roar but little inferior to that of the rapids below Niagara."[15]

Hubert Chanson has studied the acoustics of the bore near Mont Saint-Michel in northern France.[16] The rumble of the main wave is caused by bubbles in the bore roller, along with higher frequencies from the waves crashing onto rocks and bridge piers. Low frequencies between 74 and 131 hertz dominate, equivalent to a low octave on a piano.

If a writer needed adjectives to describe the sound of a tidal bore, she could do worse than consult "The Cataract of Lodore" by the Romantic poet Robert Southey. Written in the early nineteenth century, the poem depicts the Lodore Falls, a waterfall in the English Lake District, using onomatopoeia. Stretching over more than a hundred lines, it probably exhausts the lexicon of descriptors for moving water: "And whizzing and hissing... And moaning and groaning... And thundering and floundering." But water sound is more than just waterfalls and rare tidal bores; there is immense pleasure to be taken from the quiet and subtle, like a babbling brook. The remarkable thing is that in both a roaring tidal bore and a lazy winding creek, the tiny air bubbles make sound at the frequencies where our hearing works best. The physics seem just right for Southey's Romantic poetry. But maybe this is more than coincidence. Perhaps our hearing has evolved specifically to discern the frequencies produced by running water. After all, if our hearing worked in a different frequency range, we would be deaf to water, a substance vital for survival.

The frequency of the plink when a drop falls into water can be calculated from the radius of the air bubble formed. There is also a mathematical relationship between size and frequency with frozen water. During our visit to Iceland, my wife and I were on the south coast where the calving Breiðamerkurjökull glacier forms icebergs that float away on the Jökulsárlón lagoon. The haphazardly shaped blocks, looking too blue to be natural, break up and drift out to sea, or become stranded on the volcanic black beaches. Tourists make brief stops here, snapping pictures or taking a boat tour to get close to the ice, before carrying on their journey around the main ring road. We decided to camp by the lagoon. During the night, without the noise of cars and boats, we were serenaded by a tinkling sound. Small chunks of ice on the shoreline gently rocked on the lapping waves, clinking together and making rhythmic music like sleigh bells.

The frequency of the sound depended on the size of the icicles, something that Terje Isungset, a Norwegian drummer and composer, demonstrates with his ice xylophone. Many years after my trip to Iceland, at the Royal Northern College of Music in Manchester, England, I went to hear what Isungset described as "the only instruments you can drink after you've finished playing."[17] He is an archetypal Norwegian Viking, tall with rough tousled hair, and plays while wrapped in a parka. The performance was full of atmospheric and ambient sounds, evoking memories of trips to Norway.

Like a Scandinavian summer, the concert hall was cold. Even with that precaution, the musical instruments do not last long. Dressed in a large winter coat and gloves, an assistant brings out the ice trumpet or the bars of the xylophone. Once the performance has ended, the attendant quickly wraps the instruments and whisks them away to the freezer.

The ice trumpet flares outward dramatically. It is treated at the mouthpiece to prevent Terje's lips from sticking to the instrument. It has a primitive sound, like a hunting horn, and reminds me of the conch shells I once heard in Madrid. From an acoustic perspective, the material of a wind instrument is not so important if it is hard, as discussed in Chapter 4. Shell, horn, and ice may look very different, but as far as a sound wave traveling in the bore is concerned, they are similarly impervious materials. It is the shape of the outward-flaring bore and what the musician does with his lips that are most significant. Scientific measurements have shown that conch shells have exponential flares, like a French horn, creating a distinctive timbre and helping to amplify and project the sound.[18] I imagine the ice trumpet works on the same principle.

The xylophone had five bars resting on an ice trough, with their various sizes determining the different note frequencies. The bars had been cut from a frozen Norwegian lake with a chain saw, expertly carved, and then transported all the way to England. In contrast to the trumpet, the material of the xylophone is critical because the ice is actively vibrating. Once the bar is trembling, the air molecules next to the bar pick up the vibration, creating sound waves that move through the air to the listener. The air in the trough also resonates, amplifying the oscillations of the air and making the sound louder.

Terje cannot use just any old ice. He must find ice with the right microscopic structure. As Terje explains, "You can have 100 pieces of ice; they will all sound different. Perhaps three will sound fantastic."[19] The microscopic structure of a bar depends on how many impurities there were in the water when it froze, and the conditions under which the ice formed, especially the ambient temperature, which affects the speed of freezing. A slow freezing process is best because it allows the crystalline structure to form in a regular pattern with fewer flaws, enabling the ice to ring rather than emit a disappointing thud.

The ice instrument sounded like a member of the xylophone family, but I could immediately hear that its bars were not made from wood or metal. They clinked like an empty wine bottle being struck with a soft mallet. The pure, clear note suited the material perfectly. But these last two adjectives—*pure* and *clear*—might just be evidence of how aural judgments are affected by what we see. What other sound could a transparent bar make, apart from a clear one?[20]

Scientists have found that we can reliably discriminate between materials only when those materials have very different physical properties, like wood and metals.[21] Listeners latch on to how long the ringing lasts. The internal friction within a grainy wood is higher than within metal, so the wood stops vibrating sooner. This is why a rosewood xylophone makes a "bonk" and a metal glockenspiel tends to ring.

The clink of the ice xylophone was a long way from the cracking, booming, and zinging heard by the harvesters as they cut into the frozen lake to make Terje's instruments. Wait quietly by a frozen lake as the sun comes up and the ice might shift and crack, or wait as the sun sets and the ice will start to crackle and sing as it cools. These are sounds of geology in motion, an auralization of forces that shape our planet. Scientists have been measuring the noise from these seismic activities using hydrophones to estimate the thickness of ice sheets in the Arctic.[22]

To find out more about the incredible range of natural sounds from ice—the cracks, fizzes, bangs, and twangs—I met up with artist Peter Cusack in a noisy cafe in Manchester. A member of the sound intelligentsia, Peter speaks softly and describes what he hears with great precision. Peter told me about the ten days he had spent recording at Lake Baikal in Siberia. Nicknamed the "Pearl of Siberia," the lake holds about 20 percent of the world's fresh surface water, which is more than in all the North American Great Lakes combined. The thick ice sheet gradually melts in the spring, first by splitting into separate flows. Thin, icicle-shaped pieces break off from the edges of the sheet and drift about on the surrounding water, nudged by wind and waves. Millions of these ice shards jostle, creating what Peter described as a "tinkling, shimmering, hissing sound."[23]

On the opposite side of the world, at the Ross Sea in Antarctica, sound recordist Chris Watson captured a similar transformation from glacial ice to seawater using hydrophones, either underwater or wedged into glaciers. The Ross Sea is a deep bay of the Southern Ocean where early Antarctic explorers such as Scott, Shackleton, and Amundsen were based. Chris described huge blocks of ice, some the size of houses, calving from the glacier and landing on the still-frozen sea. The calving sound was explosive, like a percussive bang from a pistol. The ice also rubbed and scraped together to create "a remarkable squeaking ... sound[ing] like 1950s or early 60s electronic music."[24] Aboveground, the ice was largely silent and appeared inert, but Chris's hydrophones revealed how mobile it was beneath the surface. Later in the transformation, "Slush Puppie" ice produced grinding and crushing noises. "One of the most powerful sounds I have heard, because you realize what you are hearing," Chris explained. The Southern Ocean was moving this vast mass of ice, causing it to break up, from tens of miles away.

Walk on a thickly frozen lake, and thunderous reverberations can ricochet through the ice as it rearranges itself. On thinner ice, throwing rocks onto the surface can create alien chirps. On a winter's day, mountain biking in the Llandegla Forest in northern Wales, not long after hearing the concert of ice music, I came across a frozen reservoir with a thin layer of ice, about 5 centimeters (2 inches) thick. Skimming stones on the surface produced repeated twangs, like a laser gun from a sci-fi movie. The sounds appeared alien because each twang had a quick downward drop in pitch, a glissando that is rarely heard in everyday life.

Each time the stone struck the frozen surface, a short-lived vibration traveled through the ice before radiating into the air as a twang. In air, different sound frequencies travel with the same speed, so they all arrive at the same time. But ice is different. The high frequencies move fastest and thus arrive first, followed by the slower, lower frequencies arriving at the end of the glissando. The same effect happens in long wires. When sound designer Ben Burtt was creating effects for the *Star Wars* films, he based the laser gun on a recording of a hammer hitting a high-tension wire that was holding up an antenna tower.[25]

According to Swedish acoustician and skater Gunnar Lundmark, the chirping sound of ice can be used to test the thickness and safety of frozen lakes. As a skate moves across the surface, it creates tiny vibrations in the ice, which create a tone whose dominant frequency depends on the thickness of the frozen layer. You cannot hear the note from your own skate, because it squirts out sideways, but you can hear the sound from a friend's skate about 20 meters (65 feet) away. Lundmark did a series of measurements to test this out: "My assistant, my little lightweight son . . . hit the ice with an ax[e] and I . . . recorded the sound with a microphone and a mini-disc at a safe place."[26] He concludes that if the tone was at 440 hertz (in musical terms, the A used by orchestras to tune up), then the ice in most cases is safe, but if the tone is a bit higher in frequency—say, 660 hertz (or an E, five white notes up on a piano keyboard)—then the ice thickness is only about 5 centimeters (2 inches) and is dangerously thin. To take advantage of the singing ice, however, a skater needs to identify the frequency or equivalent musical note, which is something only people with absolute pitch can do. Unmusical skaters will have to discern ice thickness some other way.

With ice, the size of an icicle and the frequency of the noise it creates are inexorably linked. The same is true for air bubbles in water. Is there a similar mathematical relationship between grain size and frequency for singing sand dunes? One would expect so because such a relationship exists for most sound sources: violins are smaller than double basses. But whether the sand grain size is important to the frequency of the booming dunes has been hotly debated, and thus far the data have been inconclusive. However, recent laboratory tests by Simon Dagois-Bohy and colleagues at Paris Diderot University in France may have tipped the balance of scientific evidence, showing that grain size dictates the dune's frequency. Dagois-Bohy took sand from a dune near Al-Ashkharah in Oman and showed that when the sand was sieved to select a particular grain size, the boom altered. Before sieving, the grains ranged from 150 to 310 microns in size, producing a hum over a broad frequency range from 90 to 150 hertz. When the sand was sifted to select a narrower range of grains, from 200 to 250 microns, a clear single note at 90 hertz was heard.[27]

The early-twentieth-century adventurer Aimé Tschiffely once slept on a booming dune on the Peruvian coast during his 16,000-kilometer (10,000-mile) horseback ride from Argentina to Washington, DC. A report tells how the "natives" explained to him that, "the sand hill . . . was haunted and that every night the dead Indians of the 'gentilar' danced to the beating of drums. In fact, they told him so many blood-curdling stories about the hill that he began to consider himself lucky to be alive."[28] Unsurprisingly, a rich vein of folklore develops around unexplained natural sounds. Writing about rock art in North America, Campbell Grant notes the frequent drawing of thunderbirds and says, "Thunderstorms were believed to be caused by an enormous bird that made thunder by flapping its wings and lightning by opening and closing its eyes."[29]

Thunder has two distinct acoustic phases: the crash and the roll. There is an old thunder sound effect, originally recorded for the film *Frankenstein* in 1931, that perfectly encapsulates these two stages. SpongeBob SquarePants, Scooby-Doo, and Charlie Brown are among the cartoon characters that have been scared by this particular recording. Indeed, it has been used so widely that for many years, if you saw a haunted house in a storm, this is the thunder you would have heard.[30] The noise is actually quite tame, and my strongest memories of actual thunderstorms are much more scary. I can remember leaping out of bed petrified by a crack of thunder so loud that I thought my house had been struck. Hollywood sound designer Tim Gedemer explained to me that if he wants to reproduce a big thunderclap for a film—one that rips across and lights up the whole sky, one that "hits you in the gut"—it is impossible to use just a recording of thunder from nature. You might start with a real recording, but then he would add sounds that are not from thunderstorms to get a "visceral experience."[31]

As a child, I was taught to count the time between the flash of lightning and the rumble of thunder to estimate how far away an electrical storm was. The calculation exploits the fact that sound travels much slower than light. Because sound moves at about 340 meters (1,115 feet) per second, then a 3-second delay between lightning and thunder indicates that a storm is about 1 kilometer away (5 seconds would indicate a distance of 1 mile). So I have never doubted that lightning causes thunder, but surprisingly, up until the nineteenth century this causal relationship was in doubt. Aristotle, the Greek philosopher and pioneer of applying scientific methods to natural phenomena, believed that thunder was caused by the ejection of flammable vapors from clouds. Benjamin Franklin (one of the founding fathers of the United States), Roman philosopher Lucretius, and René Descartes, the French father of modern philosophy, all believed the rumble came from clouds bumping into one another. One of the reasons lightning was not proved to be the cause of thunder earlier was the difficulty of studying the phenomenon. It is impossible to predict exactly where and when lightning may be produced; scientific measurements are thus often made a long way from where the action is.

Close to the striking point there is an explosion that is among the loudest sounds created by nature. The subsequent rumble typically peaks at a bass frequency of about 100 hertz and can last for tens of seconds. The electric current of the lightning creates an immensely hot channel of ionized air, with temperatures that can exceed 30,000°C (54,000°F). This heat creates immense pressure, ten to a thousand times the size of normal atmospheric pressure, which creates a shock wave and sound.[32]

Lightning follows a jagged, tortuous path to the earth. If lightning happened in a straight line, thunder would crack but not rumble. Each kink on the crooked path—the kinks occur every 3 meters (10 feet) or so—creates a noise. Together, the noises from the kinks combine into the characteristic thunder sound. The rumble lasts a long time because the lightning path is many miles long, and it takes time for the sound to arrive from all the distributed kinks.[33]

Shock waves might also be the cause of mysterious booms heard around the world. They have colorful names: *Seneca guns* near Lake Seneca in the Catskill Mountains of New York, *mistpouffers* ("fog belches") along the coast of Belgium, and *brontidi* ("like thunder") in the Italian Apennines.[34]

In early 2012, residents of the small town of Clintonville, Wisconsin, thought they were hearing distant thunder when their houses shook and they were awakened during the night. One witness, Jolene, told the *Boston Globe*: "My husband thought it was cool, but I don't think so. This is not a joke . . . I don't know what it is, but I just want it to stop."[35] These sounds were caused by a swarm of small earthquakes, as confirmed by seismic monitoring.[36] In 1938, Charles Davison interviewed witnesses to similar moderate earthquakes, and the sounds from the quakes were variously described as the boom of a distant cannon or distant blasting, loads of falling stones, the blow of a sea wave on shore, the roll of a muffled distant drum, and an immense covey of partridges on the wing.[37]

Like UFO sightings, many of the booms can be explained by nonsupernatural reasons. In April 2012, a terrifying noise in central England was attributed to sonic booms created by a pair of Typhoon fighter jets. A helicopter pilot accidentally sent out a distress signal indicating his aircraft had been hijacked, forcing the Typhoons to break the sound barrier to quickly intercept the helicopter. When a plane moves through the air at low speed, sound waves ripple and spread out from in front and behind the plane at the speed of sound. The ripples are similar to the gentle bow and stern waves created by slow-moving boats. When the plane accelerates to the speed of sound, about 1,200 kilometers (750 miles) per hour, or faster, then the sound waves can no longer move fast enough to get out of the way. These waves combine to form a shock wave, which trails behind the aircraft in a V shape like the wake from a fast-moving boat. A plane creates a continuous sonic boom, but the wake passes over people on the ground only once. As one earwitness to the Typhoon jets reported, "It was a really loud bang and the room shook and all the wine glasses on the rack shook . . . It was weird, but didn't last long."[38] (Sometimes a double bang is heard—one caused by the wake from the nose, the other by the wake from the tail.)

A sonic boom is a weakling, however, compared to the most powerful natural sound ever experienced by humans: the 1883 eruption of Krakatoa, a volcanic island in Indonesia. As one eyewitness, Captain Sampson of the British vessel *Norham Castle*, wrote:

> I am writing this in pitch darkness. We are under a continual rain of pumice-stone and dust. So violent are the explosions that the ear-drums of over half my crew have been shattered. My last thoughts are with my dear wife. I am convinced that the day of judgement has come.[39]

Captain Sampson was only a few tens of miles from the Indonesian volcano. The power of the eruption was so great that observers on the island of Rodriguez in the middle of the Indian Ocean 5,000 kilometers (3,000 miles) away heard the explosions. Chief of police for Rodriguez, James Wallis, noted, "Several times during the night . . . reports were heard coming from the eastward, like the distant roar of heavy guns." This is a vast distance for audible sound to carry—about the same distance as separates London from Mecca in Saudi Arabia.[40] I remember news reports of the dramatic eruption of Mount St. Helens in Washington State in 1980. Had the noise of that eruption been as powerful as the sound of Krakatoa, it would have been heard all the way across the northern US, reaching Newfoundland on the east coast of Canada.

The audible blasts and explosions from Krakatoa carried an astonishing distance, but other, inaudible sounds carried even farther. Volcanic eruptions produce lots of infrasound, sound at such a low frequency that it is below the range of human hearing. (The lowest audible frequency is roughly 20 hertz.) Weather barometers around the world picked up the infrasound from Krakatoa and showed that the low-frequency waves traveled around the globe seven times, a distance of about 200,000 miles, before becoming too small to be detected.

Today, scientists monitor volcanic infrasound to help forecast and differentiate eruption types, supplementing measurements of ground vibration made using seismometers. The infrasound emitted is altered by what is happening deep inside the ground, so it gives a unique insight into the internal workings of the volcano from a safe distance.

Volcanoes make other quieter but audible sounds, include bursting bubbles, magma fragments splatting onto rocks, and gas jets hissing and roaring through vents. To experience some of these without risking life and limb near an erupting volcano, head for an active geothermal area.

Iceland is like a geological textbook writ large, with sounds that powerfully portray the forces that shape the Earth. Iceland lies on the Mid-Atlantic Ridge between the North American and Eurasian tectonic plates. The diverging plates are shaping the landscape through earthquakes and volcanic activity; the countryside is strewn with cinder cones, knobbly lava fields, and rocky rifts. At Hverir in the north, the apricot-colored landscape looks as though it is suffering from a chronic bout of yellow acne. An evil sulfurous smell invades the nostrils, and visitors have to be careful where they tread, lest they end up knee-deep in scalding liquid.

Dotted here and there are waist-high cairns of rock, gravel, and earth that hiss alarmingly as steam rushes out, sounding as though they are in imminent danger of exploding. Groundwater percolates downward for over half a mile before being heated by magma and forced back to the surface as superheated steam at 200°C (400°F). The high-speed steam escaping through gaps in the mounds jostles the adjacent still air. The result is invisible spiraling of the air that causes the hissing. You might visualize these small vortices as tiny versions of the Great Red Spot on the surface of Jupiter or a spiraling tornado.

Elsewhere, tumultuous pools of battleship-gray mud bubble at a low simmer. They almost seem alive; some belch like a thick, gloppy lentil soup, while others rage and splatter like an unappetizing, thin gruel on a fast boil. Some have an almost regular rhythm, sounding like sped-up music. Hydrogen sulfide creates both the pervasive odor and the mud within the bubbling pots as sulfuric acid dissolves the rocks.[41] The slurry in the mud pots is thrown up into the air by superheated steam and splashes into the water. Though no scientist has studied mud pot acoustics, I presume bubbles are what make the sound, as in a waterfall.

Struggling to find a scientific paper on the sound of mud pots, I contacted Tim Leighton from the University of Southampton. Tim looks like a middle-aged Harry Potter but is an expert in bubbles, not potions. He had not looked at mud pots, but he did mention the model geyser he had built at the age of twelve, complete with pressurized boiling water. Every three minutes, it erupted, throwing hot water an impressive 2–3 meters (6–10 feet) into the air. Tim lamented, "Unfortunately, I didn't know then to write the work up and submit it to a journal. But I now have a duplicate in a lab directly below my office."[42]

The word *geyser* comes from the hot spring Great Geysir in south-western Iceland. Unfortunately, Great Geysir has not erupted naturally for many decades, but close by is Strokkur (the "churn"), which produces a 30-meter-high (100-foot) spout every few minutes. Amid the crowd standing behind a rope barrier is an excited multilingual chatter as the onlookers try to guess when the geyser will erupt. The first sign is a dome of water that billows up from an opening in the ground, quivering like a giant turquoise jellyfish before, suddenly, with a whoosh, hot water is sent flying high into the air. When the water hits the ground, it fizzes and hisses, like sea waves crashing onto rocks.

Geysers are rare because they need an unusual set of conditions. Belowground, the natural plumbing must have watertight walls, with a supply of water to replenish the pipework and geothermal heat. Superheated water fills the plumbing from the bottom, while colder groundwater enters the pipework nearer the surface. The weight of the cold water at the top allows the heated water lower down to exceed the normal boiling point of water without bubbling. The dome bulges from the top of Strokkur when the plumbing is completely full. Inevitably, a few steam bubbles form that push a small amount of water out the top of the geyser. This release of water lowers the pressure deep inside, causing an explosive formation of more steam from the superheated water. The steam then forces the column of water out the top of the geyser and high into the air.[43]

Some of the most amazing natural sounds, like those generated by Strokkur, originate in places far from habitation. A few years before venturing into the Mojave Desert, I encountered tuneful sand at Whitehaven Beach on Whitsunday Island, Australia, but that was a soprano compared to Kelso's bass voice. The hot, blindingly white sands of the Australian beach squeaked at a much higher frequency, typically around 600–1,000 hertz. I came across the sound effect unexpectedly while on vacation, and I had fun shuffling up and down the beach trying to get the best squeal. Charles Darwin described hearing something similar in Brazil: "A horse walking over dry coarse sand, causes a peculiar chirping noise."[44] This high-pitched sound is more common than the booming of sand dunes, and in Australia you can even go to a place called Squeaky Beach.

The fact that a squeaky beach or a droning dune produces distinct notes you can sing along with implies a coordinated motion of the sand grains. If the grains moved haphazardly, the sound would be more like the random rustling of leaves in a deciduous tree. The opening lines of Thomas Hardy's pastoral novel *Under the Greenwood Tree* reveals how complex the sound of wind through trees can be:

> To dwellers in a wood almost every species of tree has its voice as well as its feature. At the passing of the breeze the fir-trees sob and moan no less distinctly than they rock; the holly whistles as it battles with itself; the ash hisses amid its quiverings; the beech rustles while its flat boughs rise and fall. And winter, which modifies the note of such trees as shed their leaves, does not destroy its individuality.[45]

Scientists such as Olivier Fégeant from Sweden have been researching the different ways these sounds are made.[46] For a deciduous species in leaf, such as the beech described by Hardy, the leaves and branches bang into each other as the tree sways in the wind, causing the leaves to vibrate and rustle. A birch tree is a good impersonator of waves crashing on a shore.[47] When the wind gets stronger, more impacts create a louder sound, but the dominant frequency remains surprisingly unaltered.

Fégeant is interested in whether tree rustles might hide the swish of wind turbine blades. Most wind turbines are usually very quiet, but in remote locations there are very few other sounds to mask even the tiniest murmur made by the blades. Fégeant concluded that aspen was the best deciduous choice out of the trees he tested, creating sound 8–13 decibels louder than birch or oak. With an increase in 10 decibels being roughly a doubling in perceived loudness, an aspen will sound twice as loud as the other trees. However, there is an obvious drawback to using deciduous trees, because in winter they have no leaves and thus cannot rustle.

Evergreens offer the possibility of year-round noise. At the foot of the Kelso Dunes I heard the wind whistling through the wispy foliage of the tamarisk tree. A distinct note waxed and waned, but it was not a clear sound like a musical instrument produces. It was more like the sound that a child learning to whistle would make: a tone could be heard, but it was breathy and inconsistent. The soughing is made by the movement of air around the needles, like the whistling of wind through telegraph wires. (The production of these Aeolian tones is discussed in Chapter 8.) Each needle creates a tone whose pitch depends on the wind speed and needle diameter. Thousands of these tiny sound sources conspire to create Hardy's moaning and sobbing sound. For Fégeant's measurements on spruce and pine with a moderate breeze of 6.3 meters (21 feet) per second, the soughing was at 1,600 hertz, a frequency toward the top end of the flute's range. Double the wind speed to a strong breeze, and the breathy tone increases in frequency by about an octave, to around 3,000 hertz, a note within the piccolo's range.

Hardy's description of trees moaning can be very apt, because as the wind speed naturally decreases, the sound droops in frequency like a sad person talking. For me, the tamarisk whistles were too high a frequency to remind me of moans. They were more like the sounds made by the casuarina tree (she-oak) in Australia. The drooping foliage consisting of spindly branchlets is known to create an eerie whistling, an ideal haunted-house effect for a movie. Mel Ward, a naturalist who spent many months living on islands around the Great Barrier Reef, wrote of being "lulled by the music of the sea and the sighing trees."[48] Unfortunately, nowadays the casuarina tree is largely missing from tourist destinations, and "air-conditioning, music and other amenities mask or obliterate the sounds of the outdoors." The tamarisk brought back memories of beach holidays for me. I now realize that the whooshing sound I associate with childhood visits to the seaside was made by wind whistling through the gorse bushes on the sea cliffs.

Wind can make annoying sounds when it whistles through man-made structures. Completed in 2006, the 171-meter-tall (561-foot) Beetham Tower in Manchester, England, periodically features in local headlines because it howls in the wind. Once, the hum got so loud that it disrupted the recording of *Coronation Street*, the world's longest-running soap opera. (The set for the program is located only 400 meters, or 440 yards, from the tower.)[49]

Rising vertically out of the top story of the skyscraper is a sculptural louver of glass planes supported on a metal scaffold. This structure made the tower the tallest residential building in Europe when it was finished, but it also causes the whistling. In very strong winds, the air blowing across the edge of the glass panes creates turbulence and noise. The turbulence is due to haphazard changes in air pressure—a smaller-scale version of the turbulence that causes aircraft to shudder and drop. Since sound is fundamentally tiny variations in air pressure, the turbulence creates noise (the same thing happens when a flutist blows across the instrument's mouthpiece). A temporary solution to the tower's hum, implemented in 2007, was to attach foam to the edge of the glass to mask the sharp edges and so prevent them from creating turbulence. Later that year, aluminum nosing was added, which stopped the noise for moderate wind speeds, but the building still defiantly hums in the worst storms.[50]

Turbulence is often created as wind rushes past structures such as bridges, railings, or buildings, but usually the noise is too quiet to be audible. Beetham Tower has awakened local residents and prompted dozens of complaints to the local government. To make noise that loud requires amplification by resonance. For a flute, it is the resonance of the air within the instrument that increases a note's volume. For Beetham Tower, the resonances arise from the air trapped between the many parallel rows of deep glass panes.[51]

At low wind speeds, the turbulence noise generated by the edges of the glass is below the natural resonant frequency of the structure, and the tower is quiet. This fact hints at one answer to the problem: changing the size and spacing of the glass panes so that the resonance is at the wrong frequency to be excited by high winds. Such a solution solved the whistling of New York's CitySpire Center. The drone from the tower was so bad that the building's managing agents were fined for the noise, although for only the paltry sum of $220.[52] The drone was about an octave above middle C and likened to a Second World War air-raid siren. It was generated by louvers that made a dome at the top of the building. Removing half of the louvers lowered the resonant frequency and solved the problem.

My late-night trip to hear Beetham Tower started unexpectedly. Idly browsing the web before bed, I noticed tweets from people complaining of being kept awake by the hum. One message from an acoustic engineer reported a level of 78 decibels about 100 meters (110 yards) from the base of the tower, which is equivalent to the volume you would experience while standing close to a tenor saxophone playing at a moderate volume level.[53] When I went into my garden, I could hear a faint humming. Was that the tower, the drone of a nearby road, or maybe a distant helicopter? I pulled on some clothes over my pajamas, grabbed my sound recorder, jumped into the car, and drove into town. Ignoring the cold winter air, I opened up the sunroof and drove around the city with a microphone sticking up toward the night sky recording the hum.

I immediately confirmed that the hum was the same sound I had heard in my garden, which meant that it was carrying at least 4 kilometers (2½ miles) across the city. Ironically, it was difficult to get a good recording because it was so windy. Gusting air was creating turbulence across my microphone; the same physics causing the building to sing was ruining my recording. I put a foam windshield over the top of the microphone to alleviate this problem, but in such high winds it was almost useless.

The hum came and went as the wind gusted. It was an eerie, long note from a bass musical instrument, a distinct tonal sound at 240 hertz (roughly B below middle C). Because it was distinctive, the note was easy to pick out among the traffic noise, and this is probably why residents found it so disturbing. Our hearing finds it hard to ignore such tones—sounds one can sing along with—because they might contain useful information. After all, the vowels in speech (*a, e, i, o, u*) are often pronounced in a singsong way with distinct frequencies. Knowing that tones stand out to us also explains the simple short-term solution used by the television sound recordists for *Coronation Street*. By the addition of very quiet broadband noise to the sound track—the rumble of a distant busy road would have been a good choice—the hum was hidden by a noise less likely to grab a listener's attention.

But just as the burping on the sand dunes is only the starting point of the sound, the noise created as wind whistles past the edges of the glass in Beetham Tower is only the initial impulse. Both the sand and the wind noise require amplification. For the dunes, what causes the amplification is still debated. One theory focuses on a layer of dry, loose sand, roughly 1½ meters (5 feet) deep, that sits on top of harder-packed material lower down.

Nathalie Vriend explained to me that the layer hypothesis came from her doctoral supervisor Melany Hunt, of the California Institute of Technology. To test Hunt's theory, Nathalie has carried out field measurements on various dunes in the southwestern US. To reveal the underlying structure, she turned to geophysics, using ground-penetrating radar and seismic surveys. She also described using a probe 1 centimeter (about half an inch) in diameter to take samples from the dune. Getting the probe to go through the top layer of loose sand was straightforward, but about 1.5 meters (5 feet) down it hit a layer as hard as concrete: "We had our biggest, most muscular guy hitting that probe with the hammer and he couldn't get it in any further."[54] A sample from the top of this very hard layer showed moist sand grains welded together with calcium carbonate, forming a largely impenetrable barrier to sound.

The top layer of loose sand acts as a waveguide for sound, similar to the way an optical fiber channels light. The avalanching sand produces a range of frequencies. The waveguide then picks out and amplifies a particular note. Similarly, the wind whistling past the louver of Beetham Tower creates sounds at many frequencies. The resonance between the glass panes then selectively amplifies particular notes, creating the audible hum.

Others have disputed whether a layered dune is needed, however. Simon Dagois-Bohy and colleagues re-created the booming sound in the laboratory by tipping a small sample of sand down a slope made from a thick, heavy, chipboard work surface lined with fabric. According to their theory, the sand falls in a synchronized avalanche, with grains bumping over each other at a regular rate, turning the top of the dune into a loudspeaker and producing a distinct note. But why the grains should synchronize is not known. If the theory is right, then the waveguide measured by Nathalie Vriend might just embellish the sound rather than being the underlying cause. Or maybe the waveguide aids the synchronization of the grains.

Wind plays an important role in sifting the grains of musical dunes. The mustard-colored sands at Kelso rise incongruously out of the surrounding landscape of desolate scrub and distant granite mountains. The prevailing westerly winds pick up sand from the Mojave River sink at the mouth of Afton Canyon and deposit their cargo at Kelso. Eddies form that drop sand onto the 180-meter-high (600-foot) dunes. Sand is made up mostly of grains—most commonly pieces of quartz, with smaller particles called fines. The unusual flow of winds sifts the sand so that the grains all have a similar diameter on the leeward size of the dune and there are very few fines.

The burping happens because the grains of sand are rounded and are all very similar in diameter. A grain varnish seems to be an important part of the sound production. French physicist Stéphane Douady found that his laboratory samples of sand could lose their voice. He then discovered that rinsing and drying the sand at high temperature with salt got them speaking again. The process added a varnish of silica-iron oxides to the sand, changing the friction between neighboring grains.

Diane Hope and I set off from the Kelso campsite at sunrise on day two, so that we could climb the dune when it was cooler and there was no wind. It was the summer solstice, and as we packed away the tents, a spectacular V-shaped sunbeam lit the sky through the tops of the nearby mountains.

When I checked Nathalie Vriend's scientific paper about Dumont Dunes in California, I noticed that her measurements of the boom were done over much longer slopes than I had been sliding down the day before. The paper also indicated that a steeper angle, about 30 degrees, was needed. Climbing up the dune, Diane and I scanned the hill for the longest, palest sand free of vegetation. On day one we had already learned that the sand with the gray tinge would not burp; it was easier to walk on, and the sand did not flow easily. Nearly all dunes that sing tend to do so on the leeward face, so we aimed for a ridge that was not at the top of the dune but had a long, steep slope more perpendicular to the prevailing wind than the places we had tried the day before.

With trepidation, I did a trial slide. It immediately felt different from the previous day's slope. I could feel the ground vibrating under my bottom. For a fleeting moment the sand broke into song. We had found the dune's audio sweet spot; all I needed to do now was perfect my sliding abilities. As you slide down, the sand bunches up around you. You need to avoid digging in too deep and coming to a halt, while still getting enough sand moving to create the boom.

Many writers assign a musical quality to the sound because it has a distinct frequency (88 hertz, from one of our measurements—equivalent to a low note on a cello) colored by a few harmonics. It reminded me of the drone of a taxiing propeller aircraft at an airport. The Marquess Curzon of Kedleston wrote, "First there is a faintly murmurous or wailing or moaning sound, compared sometimes to the strain of an Aeolian harp . . . Then as the vibration increases and the sound swells, we have the comparison sometimes to an organ, sometimes to the deep clangor of a bell . . . Finally, we have the rumble of distant thunder when the soil is in violent oscillation."[55] What this description misses is the whole-body experience that accompanied my triumphant slides. The drone moved my eardrums, the avalanche was vibrating my lower body, and the rest of me was quivering with excitement because I had gotten the dune to sing.

7

The Quietest Places
in the World

While on my expedition to record singing sand dunes, I experienced something quite rare: complete silence. The scorching summer heat kept visitors away; most of the time my recording companion, Diane Hope, and I were on our own. We camped at the foot of Kelso Dunes, in a barren, scrubby valley with dramatic granite hills behind us. Virtually no planes flew overhead, and only very occasionally did a distant car or freight train create noise. The conditions were wonderful for recording. No noise meant there was no need for second takes. Much of the day, however, there was a great deal of wind, which often whistled past my ears. But at twilight and early in the morning the winds calmed down, and the quiet revealed itself. Overnight I heard the silence being interrupted only once, when a pack of nearby coyotes howled like ghostly babies, unnerving me with their near-musical whistling and chattering.

High up on the dune, early on the second morning, I was waiting for Diane to set up some recording equipment. Since she was some distance away, I had a chance to contemplate real silence. The ear is exquisitely sensitive. When perceiving the quietest murmur, the tiny bones of the middle ear, which transmit sound from the eardrum to the inner ear, vibrate by less than a thousandth of the diameter of a hydrogen atom.[1] Even in silence, tiny vibrations of molecules move different parts of the auditory apparatus. These constant movements have nothing to do with sound; they stem from random molecular motion. If the human ear were any more sensitive, it would not hear more sounds from outside; instead, it would just hear the hiss generated by the thermal agitation of the eardrum, the stapes bone of the middle ear, and the hair cells in the cochlea.

On the dunes, I could hear a high-pitched sound. It was barely audible, but I worried that I might be experiencing *tinnitus*—that is, ringing in the ears, perhaps evidence of hearing damage caused by my excessively loud saxophone playing. Medics define tinnitus as perceiving sound when there is no external source. For 5–15 percent of the population tinnitus is constant, and for 1–3 percent of people it leads to sleepless nights, impaired performance at tasks, and distress.[2]

Theories of tinnitus abound, but most experts agree that it is caused by some sort of neural reorganization triggered by diminished input from outside sounds. Hair cells within the inner ear turn vibrations into electrical signals, which then travel up the auditory nerve into the brain. But this is not a one-way street; electrical pulses flow in both directions, with the brain sending signals back down to change how the inner ear responds. In a silent place, or when hearing is damaged, auditory neurons in the brain stem increase the amplification of the signals from the auditory nerve to compensate for the lack of external sound. As an unwanted side effect, spontaneous activity in the auditory nerve fibers increases, leading to neural noise, which is perceived as a whistle, hiss, or hum.[3] Maybe what I was hearing on the dunes was the idling noise

Figure 7.1 The anechoic chamber at the University of Salford.

of my brain while it searched in vain for sounds. One thing I noticed was that this high-frequency whistle was not always there—maybe a sign that, after a while, my brain habituated to the noise.

In contrast to the variable silence on the dunes, at my university there is an *anechoic chamber*, a room that provides unchanging, guaranteed silence, uninterrupted by wind, animals, or human noise (Figure 7.1). The anechoic chamber never fails to impress visitors, even though the entrance is utilitarian and uninspiring. Just outside the entrance they see dusty metal walkways, and nearby, builders are often making lots of noise constructing test walls in a neighboring laboratory. These walls will be analyzed for how well they keep sound from passing through them. Guarding the anechoic chamber are heavy, gray, metal doors. In fact, you have to go through three doors to reach the chamber, because it is a room within a room. To make the place silent, several sets of heavy walls insulate the innermost room, stopping outside noise from entering.

Like a modern concert hall, the chamber is mounted on springs to prevent unwanted vibration from getting into the inner sanctum.

The chamber is the size of a palatial office. First-time visitors are usually circumspect, not least because the wire floor is like a taut trampoline. Once inside, with the doors closed, they notice vast wedges of gray foam covering every surface, including the floor beneath the wire trampoline. When showing visitors around, I like to say nothing at this point because it is fun to watch the realization sweep across their faces as they adjust to this unbelievably quiet space.

But it is not silent. Your body makes internal noises that the room cannot dampen. Sound recordist Chris Watson described his experience in such a chamber: "There was a hissing in my ears and a low pulsing that I can only guess was the sound of my blood circulating."[4] The internal noises are not the only oddity. The foam wedges on the floor, ceiling, and walls absorb all speech; there are no acoustic reflections. We are used to hearing sound bouncing off surfaces—floor, walls, and ceiling—which is why a bathroom is lively and reverberant, and a bedroom muffled and subdued. In the anechoic chamber, speech sounds very muffled, like when your ears need to pop in an airplane.

According to the *Guinness Book of Records*, the anechoic chamber at Orfield Laboratories in Minneapolis is the quietest place in the world, with a background noise reading of −9.4 decibels.[5] But how quiet is that? If you chatted with someone, your speech would measure around 60 decibels on a sound-level meter. If you stood quietly on your own in a modern concert hall, the meter would drop down to a level of about 15 decibels. The threshold of hearing, the quietest sound a young adult can hear, is about 0 decibels. The test room at Orfield Laboratories, like the chamber at Salford University, is quieter than that.

An anechoic chamber has an impressive silence because it simultaneously presents two unusual sensations: not only is there no external sound, but the room puts your senses out of kilter. Through their eyes, visitors obviously see a room, but they hear nothing that indicates a room. Add the claustrophobic drama of being enclosed behind three heavy doors, and some begin to feel uneasy and ask to leave. Others are struck with fascination at the oddness of the experience. I know of no other architectural acoustic space that regularly has such a strong effect on people. But it is remarkable how quickly the brain gets used to the silence and the contradictory messages from the senses. The exotic sensory experience is filed in memory, and the extraordinary becomes more normal. The magical impact of the first visit to an anechoic chamber is never really experienced again. Not only are anechoic chambers very rare, but our brains ensure that the experience is mostly ephemeral.

However, there is more to silence than experiencing the quietest rooms on Earth. Silence can be spiritual; it can even have an aesthetic and artistic quality, as epitomized in John Cage's famous silent composition 4'33". When one of my teenage sons learned I was going to see this piece performed, he expressed shock that I would spend money to hear nothing. Cage composed the piece in 1952 after a visit to the anechoic chamber at Harvard University. There, surround by thousands of fiberglass wedges, he had expected to find silence. But it was not entirely quiet, because of the noises within his own body. He also described hearing a high-frequency sound that might very well have been caused by tinnitus.

The performance of 4'33" that I heard took place nine months before my trip into the desert. It was carried out with all the usual pomp and ceremony of a normal concert. The house lights were dimmed, and the musician strode on stage and bowed to the applause from the audience. He then sat down at the piano, adjusted the seat to make it just the right height, turned the page on the score, opened the keyboard lid of the piano, closed it again, and started a timer. Nothing else happened, apart from the occasional turning over of the empty sheet music, and the opening and closing of the keyboard lid to signify the end and beginning of the three movements. At the end, the pianist opened the keyboard lid for one last time, stood up to accept the applause of the audience, bowed, and left. Amusingly, the work comes in different orchestrations, and I guess the full orchestral version is very popular with the Musicians' Union, maximizing the number of people being paid to play no notes.

The first surprise occurred before the pianist entered the stage. As the doors of the auditorium were closed and the house lights dimmed, I felt a sudden frisson of excitement, even greater than I get before a normal concert. A modern concert hall is one of the quietest places to be found in a city. At the Bridgewater Hall in Manchester, England, tour guides like to recount the story that when the largest peacetime bomb ever detonated in Great Britain exploded in 1996, workers within the auditorium did not hear the bang, because the hall was so well isolated from the outside world. Planted by the Irish Republican Army (IRA) in the city center, the bomb destroyed shops, broke virtually every window within a kilometer (half-mile) radius, and left a 5-meter-wide (16-foot) crater.

It is worth taking the backstage tour of a modern concert hall to see the precision needed to achieve the noise isolation. The tour guides are usually very proud of the fact that the auditorium is built on springs. Like a souped-up car suspension, the springs stop vibration from entering the concert hall. If ground vibration were to set parts of the auditorium moving, the tiny vibrations of the hall would set air molecules into motion, creating audible noise. Everything connected to the hall that might transmit vibration—electricity cables, pipes, and ventilation ducts—have to be carefully designed with their own little suspension systems. The attention to detail is staggering.

In recent decades, classical concert halls have been built to be quieter and quieter, giving conductors and musicians access to the widest possible dynamic range to exploit and create drama. In a good modern hall, the collective noise from audience members breathing and shuffling in their seats is actually louder than any background sounds from outside noises or ventilation systems.[6]

What the audience hears during a performance of *4'33"* depends on the isolation of the auditorium and the quietness of the audience. The hall I was in did not have the best sound insulation, and I could occasionally hear buses on the busy road outside. The audience was small, about fifty people, whom I could hear fidgeting and coughing. With these distractions, as the piece progressed I found my mind wandering. But were these really distractions, or the actual music? Although there was a musician on the stage, what Cage's piece does is shift the focus from the performers to the audience. And that change from being a passive member of the audience to being part of the performance was at the heart of my second surprise. When the piece was over, I felt a strong sense of a communal achievement with everyone else in the audience and the performer. As the audience clapped and a few shouted "More!" and "Encore!" I had an overwhelming sense of a shared experience. We had all just done something that was completely pointless—or was it?

Moments of silence are commonly used in the arts—famously so in theater by playwrights Harold Pinter and Samuel Beckett. For Pinter, silence forces the audience to contemplate what the character is thinking. For Beckett, silences might symbolize the meaningless and eternity of existence.[7] Short silences are also used regularly in music. A jazz group in full flow may stop abruptly for a moment, before resuming exactly together a couple of beats later and carrying on as though the pause had never happened. The silence adds dramatic tension by subverting expectation in a way that the brain finds pleasurable.

Imagine a musician walking up to a piano and repeating a snippet of a favorite melody over and over again. The predictability would soon become tedious. Similarly, one would derive little pleasure from the more random approach of letting a cat run about on the keys playing notes haphazardly. Successful music is neither completely repetitive nor entirely random. It lies somewhere between, having some regular rhythmic and melodic structure, but with changes to maintain the listener's interest.

One task the brain does when listening to music is try to break down the rhythmic structure, the beat or groove. The seemingly simple task of finding and tapping along to a beat involves several brain regions and is not fully understood. The basal ganglia buried deep inside the cerebrum seem to play a role, as do the prefrontal cortex at the front of the brain and other areas used for processing sound.[8] The basal ganglia play a vital role in initiating and regulating motor commands; when they are damaged in Parkinson's disease, patients have difficulty starting movements.

As the brain decodes the information bombarding it during a tune, it is constantly attempting to predict when the next strong beat will occur. It draws on past experiences of similar music, and recent notes from the piece, to work out where the rhythm is going. Correctly anticipating the next strong beat is satisfying, but there is also a delight in hearing skillful musicians violate that regular tempo, subverting the listener's expectation. One way of flouting expectation is to add unexpected silences, even very brief ones. The brain seems to find pleasure in adjusting itself to remain synchronized with the musical beat.[9]

A sudden pause in music also transfers the responsibility for the beat to the audience, because for a moment they have to carry the tempo until the musicians resume playing. Like John Cage's work, the pause takes the focus of the music making away from the stage. The second piece in the concert that featured *4'33"* was a more conventional piano sonata by Charles Ives that required no audience participation. As the pianist raced his fingers up and down the keyboard, he seemed to be trying to make up for the lack of notes in Cage's work. The piece left me entirely cold, and I kept wishing I could hear silence again.

Sound mixers generally avoid complete silence in film soundtracks, with one famous exception. In *2001: A Space Odyssey*, Stanley Kubrick boldly used lots of quiet. If a film director attempted this nowadays, it would be the film equivalent of *4'33"*, and all you would hear would be endless crunching and slurping of junk food and soda by fellow cinemagoers. Often when the audience thinks there is silence, there are actually quite a few audio tracks of "nothing" playing. Charles Deenen, head of audio for Electronic Arts, described to me how he became obsessed with silent rooms when developing a video game soundtrack. Turning up the volume on recordings that he had made in empty rooms revealed "amazing creepy tones" and "amazing squeaking things happening."[10] Charles also described how he might take a sound, like a camel moan, and digitally manipulate it, shifting it down many octaves and listening for distinctive tones or ringing that might appear and create the right creepiness. Game players or a film audience might not be consciously aware of these background sounds, but they are an important part of setting the emotional feel of a scene.

"Space, the final frontier," announces James T. Kirk at the start of the first *Star Trek* episode. As the spaceship *Enterprise* flies past the screen, the voice sounds as though it was recorded in a very reverberant cathedral. I know space is a big place, but where are the reflections meant to be coming from? And anyway, space is silent or, to quote the catchy tag line from the 1979 movie *Alien*, "in space, no-one can hear you scream." For an astronaut unfortunate enough to be caught outside the spaceship without a space suit, screaming to occupy the moments before asphyxiation would be pointless, as there are no air molecules to carry the sound waves.

But Hollywood does not let anything as trivial as physics get in the way of a compelling soundtrack. The latest *Star Trek* film showed the outside of the soaring *Enterprise* accompanied by lots of powerful engine noises; the photon torpedoes sounded pretty impressive as well.

When I think of the inside of a real spacecraft, I picture people floating serenely and gracefully in zero gravity. I met NASA astronaut Ron Garan in early 2012, when he had just returned from a six-month mission on board the International Space Station. He explained to me that the sonic environment in a real spacecraft is a long way from being serene. Even outside on a spacewalk (his previous mission had included a walk that lasted six and a half hours), there is no silence. Indeed, it would have been worrying if there had been, because it would have meant that the pumps circulating air for him to breathe had stopped working. Spacecraft are full of noisy mechanical devices, such as refrigerators, air-conditioning units, and fans. Theoretically, the noise could be reduced, but quieter, heavier machines would be expensive to lift into orbit.

Studies on a single space shuttle flight found temporary partial deafness in the crew. Inside the International Space Station (ISS) it is so loud that some fear for the astronauts' hearing.[11] At its worst, the noise level in sleep stations was about the same as in a very noisy office (65 decibels). An article in *New Scientist* reported, "Astronauts on the ISS used to have to wear ear plugs all day, but are now only [required to] wear them for 2 to 3 hours per work day."[12] The need for earplugs, even for part of the day, indicates how hostile the soundscape is. Squidgy foam earplugs can reduce sound by about 20–30 decibels. The higher levels of carbon dioxide and atmospheric contaminants that exist at zero gravity in spacecraft might also make the inner ear more susceptible to noise damage.

Outer space might be devoid of audible sound, but that is not true of other planets, and scientists have put microphones on spacecraft such as the *Huygens* probe to Saturn's moon Titan to record it.

As long as a planet or moon has an atmosphere—some gas clinging to the planet—there is sound. Microphones have the advantage of being light, needing little power, and being able to hear things hidden from cameras. Mind you, the audio recorded from Titan as the *Huygens* probe descended through the atmosphere is not very otherworldly. It reminded me of wind rushing by an open car window while driving on a highway. However, when I consider where it was recorded, almost a billion miles away from Earth, this mundane sound feels much more exciting.

If a pipe organ were taken to Mars for a performance of Bach's Toccata and Fugue in D Minor, the astronauts would find the notes coming out of their musical instruments at a lower frequency. The atmosphere of Mars would transpose the music to roughly G-sharp minor. The frequency of the note produced by an organ pipe depends on the time it takes sound to travel up and down the length of the tube. Because Mars has a thin, cold atmosphere of carbon dioxide and nitrogen, sound moves at about two-thirds the speed it does on Earth. The slower round-trip up and down the organ pipe produces a lower frequency. Given the toxic gases in the atmosphere, visiting astronauts would not be taking their helmets off to sing. But if someone did dare to do this, the voice would drop in pitch like the organ pipe, turning tenors into Barry White soundalikes. Unfortunately, the sexy voice would not carry very far, because Mars's thin atmosphere is almost a vacuum.

Venus has a very dense atmosphere, which would slow down the vibration of the astronaut's vocal folds (vocal cords) and lower the pitch of the voice. However, sound travels faster in the Venusian atmosphere, which would make the resonances of the mouth and throat rise. As a result, the astronaut's voice would sound squeaky, as happens when someone talks after breathing helium. Tim Leighton, from the University of Southampton, suggests that together, these effects would make an astronaut sound like a bass Smurf.[13]

Although sound levels in the International Space Station have been reduced enough that they probably no longer pose a risk to hearing, noise can affect health in other ways. And it is not just astronauts who should be concerned. For example, someone whose sleep is disturbed by airplane noise is more likely to be tired, irritable, and less effective at work the next day. If we are exposed to high levels of noise, our bodies will produce more stress hormones in the long term that might elevate blood pressure and increase the risk of heart disease.[14] Removing noise is therefore good for us, but is a strong dose of quiet better? Should we be seeking out complete silence?

One day at the office—that is, our anechoic chamber at Salford University—as the BBC tried to measure the footfall of a centipede, sound recordist Chris Watson suggested I should visit a flotation tank, a dark isolation chamber where you float in very salty water and experience sensory deprivation. And when better to experience that than a few days after hearing silence in the Mojave Desert? I headed to Venice Beach, a bohemian district of Los Angeles famous for its skimpily dressed in-line skaters, outrageous street performers, and kooky poseurs. My appointment was after dark, when the place feels less free-spirited and distinctly more dangerous.

I went to a tatty, closed shopping mall, and the attendant opened up the shutters to let me in. The tank was in a small shop at the back. He showed me around, giving detailed instructions on what to do, before asking me to sign a lengthy disclaimer. He then announced that I could stay as long as I wanted because he was leaving. I was told to let myself out, making sure I closed the shop door securely behind me. This was unnerving. What would happen if I fell asleep? What if I could not get out? Was I going to be stuck in a flotation tank overnight? With trepidation I got undressed, stuck in my earplugs, showered, and wandered over to the tank.

From the outside, the chamber looked like a giant industrial fridge, made of metal to keep noise out, about 2.5 meters (8 feet) long, 2 meters (7 feet) high, and 1.5 meters (5 feet) wide. I climbed in, shut the door, and lay down in the body-temperature, shallow, salty water. The buoyancy of the brine kept me well afloat, but the angles between my head, neck, and back felt unnatural, and it took time for me to get comfortable. It was pitch-black; it did not matter whether my eyes were open or closed, there was nothing to see. Lying naked in the dark, unable to hear external sounds, and in a closed, rundown shopping mall, worrying thoughts besieged me. Was the attendant a new age version of Sweeney Todd?

I turned my mind to more pleasant things and tried to relax into the experience. Provided I lay still and did not splash about, I could perceive nothing from outside. I could hear the internal high-pitched whine I had noticed in the desert, but after a while it disappeared, returning only intermittently. There was also a low-frequency, pulsating sound that sometimes seemed to make my head wobble. This was pulsatile tinnitus, where the sensitive hearing system picks up the rhythmic pumping of blood. It's similar to the sensation you get when the heart starts pounding during intensive exercise. Normally, this blood movement is quieter than the everyday external sounds passing down the ear canal, but in the flotation tank with earplugs, this pulsing sign of life became audible. I heard it only occasionally, and most of the time I could hear nothing at all. To appreciate this total silence, I had to shut up the voice in my head and stop myself from listening for sounds. This is not easy to do, because the brain constantly directs attention in anticipation of hearing something. In a neuroimaging study, Julien Voisin and colleagues found increased activity in the auditory cortex during the silence before a sound was heard.[15]

The effect of cutting off both hearing and vision, and the feeling of the hot, salty water on my skin, made me very aware of touch. After a while I got the impression that my legs and arms had disappeared and my feet and hands were detached from my body. They felt slightly numb, like the sensation you get just before pins-and-needles sets in. It is hard to articulate the sound experience because it was all about an absence of hearing. This was probably the nearest to a truly perceived silence I have heard, because for long periods my sense of hearing seemed to be completely missing, to the point that my only sense seemed to be touch.

I decided it was time to finish; I struggled to my feet and groped around for the door handle. Outside of the chamber I checked my watch and was flabbergasted that I had been floating for two hours! I showered, dressed, let myself out of the shop, and sat in my car feeling very weak and queasy, probably because I was badly dehydrated. The flotation tank is supposed to help with stress management by lowering levels of cortisol, but in my shaky state I was not at all sure it had worked for me.[16]

Going to the countryside and getting some peace and quiet is meant to be good for us. But rural places are usually far from silent. Within a crowded and intensely farmed country, it is difficult to escape the sounds of agriculture and human activity. Media stories about newcomers to the countryside complaining about noise appear each year like perennial weeds:

> The mayor of a French village has forbidden complaints about farmyard noises, in a pre-emptive strike against a growing number of urban newcomers prepared to sue for their "right" to rural peace. City dwellers intending to join the 300 inhabitants of Cesny-aux-Vignes, 12 miles from Caen in Normandy, have been asked to cohabit with crowing cockerels, braying donkeys or chiming church bells "without complaint."[17]

But those people aren't alone in imagining the ideal countryside through rose-tinted ears.[18] I conjure up pastoral sounds: the bleating of sheep in fields, the trickle of water running down a stream, and the thwack of leather on willow at a village cricket match. I am not a particularly nostalgic person but, rather startlingly, I have just painted a scene from a P. G. Wodehouse novel, tales of England from a hundred years ago, with the bumbling aristocrat Bertie Wooster and astute butler Jeeves.

Take a moment to think of your ideal rural soundscape; what would you like to hear? I would be surprised if you picked complete silence, because most people go to the countryside to feel connected with nature. Gordon Hempton, a sound recordist and acoustic ecologist from Washington State, has been campaigning for the preservation of natural silence, which "is as necessary and essential as species preservation, habitat restoration, toxic waste cleanup, and carbon dioxide reduction."[19]

Gordon Hempton claims that there are very few remaining quiet places in the US, even though the country has large tracts of empty land. True freedom from man-made sounds is surprisingly hard to achieve because of the web of flight paths that crisscross the sky. Hempton has named one niche that is free from aircraft noise as "One Square Inch of Silence" and says this "is the quietest place in the United States."[20] It is located in the Hoh Rainforest, in Olympic National Park, Washington State. But there is not a complete absence of sound. Although the location is free from man-made noises, there are plenty of natural sounds to listen to in this enchanting rainforest. The lush, green canopy of ancient coniferous and deciduous trees with mosses and ferns blanketing the surfaces is home to noisy animals and birds, and the high level of rainfall produces lots of river noises. Imagine if it were truly silent, so the rapid staccato notes of the winter wren were missing and no Douglas squirrel was crying "pillillooeet." This would be a barren and lifeless place.

Murray Schafer, the pioneering acoustic ecologist who evangelizes about ear cleaning, praises rural settings and calls them a "hi-fi soundscape." A good-quality audio system reproduces sound with little or no unwanted noise. This fact led Schafer to define a hi-fi soundscape as being a place where the hearing system is not overwhelmed by undesirable noise, making useful, relatively subtle sonic information more audible. In contrast, he defined a lo-fi soundscape as one where individual sounds are masked by the rumble of traffic and other man-made noises.[21]

The US National Park Service has a policy that states, "The Service will restore to the natural condition wherever possible those park soundscapes that have become degraded by unnatural sounds (noise), and will protect natural soundscapes from unacceptable impacts."[22] The Campaign to Protect Rural England (CPRE) claims that half the people visiting the countryside do it to find tranquillity.[23] Access to tranquillity has been shown to reduce stress.[24] (The three competing theories as to why natural sounds might be good for us were outlined in Chapter 3.) Research commissioned by the CPRE found that seeing a natural landscape, hearing birdsong, and seeing the stars were the top three things contributing to a sense of tranquillity. Unwanted features included hearing constant traffic noise, seeing lots of people, and urban development. As these findings show, tranquillity is not just about sound; it includes being calm and free of disturbance, including consideration of how a place looks. It requires our senses to be in harmony, and not to have competing and conflicting stimuli.

In science, the different senses are often studied in isolation, but our brains do not take such a demarcated approach. Although we sometimes use different regions of our brains to process and interpret signals from our various senses, the overall emotional response comes from an aggregation of what we see, hear, smell, taste, and touch. Michael Hunter, from the University of Sheffield, and collaborators have shown how the brain handles sensory inputs in tranquil and nontranquil places using a functional magnetic resonance imaging (fMRI) scanner.[25] Ingeniously, they used an ambiguous sound recording in all the tests (waves crashing on a beach sound remarkably like light road traffic) and just changed the picture to make people think they were listening to something different. The natural scene of the beach increased connectivity between the auditory cortex and other parts of the brain. The increase in connectivity did not happen with the man-made highway. The results show that what we are looking at affects which neural pathways are used to process sound. When we're evaluating tranquillity, we must consider sound and vision together.

The author Sara Maitland has gone to great lengths to find tranquillity and solitude: "Chosen silence can be creative and generate self-knowledge, integration and profound joy."[26] She moved to a remote cottage, cut off communication with others, and removed the TV, clothes dryer, and as many other noisy appliances as she could. Maitland writes about the sense of spirituality that her silent existence has given her. Others have also referred to rural tranquillity in hushed, almost reverential language. Indeed, surveys have linked quiet, natural places and a sense of spirituality.

A tranquil soundscape has a quality and emotional connection similar to that found in churches: people become very sensitive to the sounds around them, but not in a stressful way.[27] Maybe this sense of spirituality simply reflects the reduced cognitive load on the brain, which experiences less stress when processing the calmer soundscape. To keep our hearing open to possible signs of danger, our brains have to constantly work to suppress unchanging noises, such as the perpetual drone of traffic. This situation is not conducive to finding a relaxed and spiritual sense of well-being.

The CPRE has even quantified tranquillity, publishing brightly colored, blotchy maps of England that define areas by how tranquil they are. Researchers calculated a tranquillity index by working out what people might see by measuring lines of sight to natural and man-made features, and also what they might hear through predictions of noise levels from roads and aircraft.[28] When I looked up my home city on the map, I found it to be a sea of red, signifying no tranquillity. Then my eye was drawn toward dark green areas farther north, just below the Scottish border, indicating large tracts of tranquil countryside.

Somewhere in the green areas was the most tranquil place in England, and I decided I wanted to visit it. But when the original research had been published a few years earlier, the CPRE had been coy about the exact coordinates of the most tranquil place, not wishing it to be ruined by visitors. So I was surprised and delighted when I was granted access to the original mapping data so that I could pinpoint the location just outside Northumberland National Park, near Kielder Forest.

The place would not be easy to reach, because by definition it was far from buildings, infrastructure, and roads. A few months after returning from the desert, I sorted out the logistics and started by cycling as close as possible on the nearest road. It was the start of autumn, and I was either too cold in the deep shadows cast by conifer forests, or out of breath and too hot climbing up hills in full sun. I traveled through countryside typical of northern England: rolling hills with fields being grazed by sheep and cattle bordered by dry stone walls. As I gained more height (the most tranquil place is almost at the top of a hill), I entered scrubby moorland with army tanks in the middle of a firing range. I now realized why this place has very little human infrastructure on civilian maps. Strange that the place rated as being very tranquil is where the gunners of fighter planes are often trained.

I turned my mountain bike off the road and onto forest trails, trying to get as close as possible to the most tranquil spot. I parked my bike, put on walking boots, and started the long trudge cross-country. As I left the trees behind, I entered one of the Kielder Mires, the local name for peat bogs covered in moss and heather. The ground was very uneven, and my feet kept disappearing into dips and gullies and getting soaked. I had previously thought of asking the CPRE if I could publish the location of the quietest place. But I now realized this would be a bad idea. If lots of people put this on their rambling bucket list, then this delicate wetland terrain would be damaged.

Fortunately, no one was logging in the forest, and the military were having a day off. It really was incredibly quiet, with the only sound coming from my pounding heart, heavy breathing, and rhythmic squelching from my boots. After an hour I judged I had reached the final location, and I turned on my phone to check the GPS coordinates. A beep announced the arrival of a text message. There was no other sound, no visible sign of human activity, and yet I had decent mobile phone reception!

To capture tranquillity, I decided to try my sound-recording equipment. Even with the gain on my recorder turned up to its highest setting, there was nothing to be captured apart from the dull background hiss of electrical noise from within the equipment, and the occasional flapping sound as I tried to kill the midges that were feasting on me. Just then, a few birds flew by in the distance, called out using rapid staccato chirps, and quickly disappeared before I could spot what they were.

This was not a very serene, rejuvenating, or relaxing silence. I was exhausted, my feet were drenched, and I was unable to sit down in the wet peat bog. The mild anxiety that the military might turn up and arrest me for trespassing did not help. But I was surprised and impressed that I had found complete silence in the English countryside. For me, the lack of animal sounds detracted from the experience, reminding me that the surrounding monoculture of conifer trees is not good for biodiversity. Complete silence in a natural setting is not necessarily wonderful. I wanted to hear birds or the trickle of a stream; even a buzzing fly would have been good—some sound that signified life.

More than half the world's population now lives in cities. Is it possible to find some form of tranquillity in an urban environment? Engineers have been working hard to make cars quieter, but with more traffic on the road, average noise levels in cities have stayed the same.[29] And considering only an average of the noise level overlooks a crucial trend. As motorists try to avoid traffic by leaving for work before or after rush hour, their noise pollutes peaceful times of the day. As drivers try to avoid gridlock, they turn calm back roads into rat runs, ruining quiet places. Although cities thrive on human activity, vibrancy, and excitement, people need relatively quiet places for respite from the hustle and bustle.

Policy makers are interested in how tranquil refuges can be preserved, but achieving such protection is turning out to be difficult.[30] The ideal would be to craft a simple metric that could be measured on a sound-level meter or predicted in a computer model. A scientific report once suggested that areas with sound levels below 55 decibels (a level you might hear from a cheap refrigerator) should be designated quiet areas; another, that man-made sounds should be below 42 decibels (a typical level for a library).[31] By those criteria, there are no tranquil areas in major cities like London, which is nonsense. Like all world capitals, London is a noisy place, but turn a corner and go down a back street and often you find a quiet square where noise is distant and less intrusive. This just illustrates the problem of trying to reduce human perceptions to simple numbers.

In a city, what matters is relative quiet rather than absolute loudness. As in the countryside, man-made sounds need suppressing, but they do not have to be inaudible. Songbirds, rustling leaves, and moving water need encouraging, because studies have shown that tranquillity in towns is greater when natural sounds are louder.

Other senses need to be considered: research suggests that places with more hard landscaping need to be quieter than greener refuges, and certain smells, especially the stench of street urination, do not aid tranquillity.

How might you make such an acoustic oasis? Layout is very important because a noise source out of sight is usually quieter. The piazza in front of the British Library in London is an interesting example. Facing onto a very busy street, it is still possible to find some quiet in this pedestrianized square because a high wall hides the road. Unfortunately, bass sound carries over the wall more easily than higher frequencies, so the rumble of waiting buses is still overpowering from time to time, but a higher wall placed closer to the road could solve this problem.

In quiet back streets, it is actually the buildings that often act as barriers to noise. Hildegard Westerkamp, a world-renowned composer, radio artist, and sound ecologist from Canada explained to me on the London soundwalk that this "quiet stone sound of back streets" is heard only in old cities where roads are narrow and buildings close together.[32] In North America, where most of the streets are wide, it is also hard to escape the hum and whine of ventilation fans. Since lowering sound levels at the source is the most effective strategy, having fewer, slower cars is good, as is reducing rolling noise through better asphalt and tire design.

As for promoting desirable sounds, water features and fountains create likable babble and splash that can also mask unwanted traffic noise. Hong Kong is incredibly crowded and noisy. Even so, close to the center there is a park with a giant aviary where man-made sounds are completely absent and the songs of the birds can be enjoyed. The traffic noise is inaudible because the babbling of a stream helps hide the sound of the roads. Sheffield in northern England was at the heart of the UK's steel industry and famed for its high-quality cutlery. Nowadays, it may be better known as

Figure 7.2 Part of *The Cutting Edge*, a fountain by Chris Knight of Si Applied,
acting as a noise barrier outside the Sheffield railway station.

the setting of the film *The Full Monty*, a black comedy about former steelworkers, unemployed and desperate for money, forming a
striptease act. At Sheffield train station there is an enormous fountain (Figure 7.2). It is on a size and scale that I expect to see only on
the grounds of a grand, stately home. I travel to Sheffield often and
had noticed this huge water feature being built, but I did not realize
the subtleties of the acoustic design until they were pointed out to
me by Jian Kang, a soundscape guru from one of the local universi-
ties. The high-sided walls of shimmering water serve as a noise barrier shielding the square from traffic. In addition, a series of large
pools stretches downhill with water flowing between them. If you
stand in the right place, the waterfalls between the pools imitate

the chuff-chuff sound of steam trains because the water stops and starts. This irregular flow draws more attention because it is harder to tune out intermittent sounds than continuous ones. The fountain makes the traffic less perceptible by both physically blocking the sound and generating pleasant, distracting water sounds.

I talked to a colleague from Salford University, Bill Davies, about the large research project he had conducted investigating positive sound design in cities. Appropriately, he is the most soft-spoken person I have ever met—so quiet that his speech becomes almost inaudible at times. Bill and his collaborators took people on sound-walks and asked them about their impressions of city squares; they also played sounds to subjects in the laboratory and asked which ones they preferred. The results showed that vibrancy and pleas-antness are important.[33] A busy piazza can be buzzing with people, yet there is a pleasant calmness if you are slightly distant—say, in a café on the edge of the square watching the crowd go by. This may be true even if cars are present. In contrast, a city square that is little more than a traffic rotary lacks the babble of the crowds. The noise from cars is just unchanging and unpleasant.

Researchers have demonstrated the health benefits of natural sounds, but I think Bill's finding on vibrancy and pleasantness indi-cates that scientists are overlooking an important sound that could also be good for us. Perhaps hearing human activity reduces stress. The quiet chatter of people in a café is relaxing and not too alert-ing. Furthermore, one should have a positive emotional response to being surrounded by other people in a friendly atmosphere. After all, being a social animal has been a vital part of our evolutionary success. Maybe this is an avenue for future research.

One thing is certain: perception of silence is highly subjective. In Kielder Forest I felt there should have been natural sounds, and the lack of them made the place seem barren. On the Mojave dunes, the enveloping silence seemed appropriate and peaceful. Looking out over the vast, barren valley, I could imagine the complete quiet extending for miles. The silence waxed and waned as the wind came and went, creating a gentle whooshing past my ears, or the occasional insect buzzed, or I heard the beating sound of bird wings. These aural accents made the silence seem very natural.

An ex-colleague of mine, Stuart Bradley from Auckland University, has visited Antarctica, another place devoid of vegetation where silence can be heard. Stuart is a tall New Zealander, sporting a fine mustache like a soccer player from the 1970s. Ironically, what Stuart does in Antarctica is make noise and briefly ruin the pristine natural soundscape. He uses a sodar (a sound radar system) to measure weather conditions, sending up strange chirps that bounce off of turbulent air in the atmosphere before returning to the ground to be measured. I asked Stuart if he had experienced silence in Antarctica, and he told me about his time in the dry valleys, possibly the most barren places on Earth, which lack snow and ice cover: "Sitting up on the valley wall on a still day, there was no sound I could identify (except heartbeat? breathing?). No life (apart from me). So no leaves either. No running water. No wind noise. I was certainly struck by the primeval 'feel.' "[34] Stuart commented on how different this was to the sound of a silent laboratory, "I didn't get the claustrophobic feel one can get in an anechoic chamber . . . I suspect this is because, although it was incredibly quiet, it was also a very, very open vista (the valley walls were 1,500–2,000 meters [5,000–6,500 feet] high and the visibility was amazing!)."

Getting away from normal existence and civilization is an important part of silent retreats. John Drever, the acoustic ecologist who had taken me to hear bitterns (see Chapter 3), suggested that I needed to experience a retreat to truly understand silence. So I signed up for three days of Buddhism in an eighteenth-century manor house out in the English countryside for a month before I went into the desert. Only when I arrived did I realize that this Buddhist weekend would involve fifteen meditations a day, challenging my inflexible, middle-aged body to adopt an unfamiliar posture for hours. It was a great relief when the gong sounded to end each backaching session. I needed to have done some training to be better prepared for the static gymnastics.

Just before silence started on the first day, we were asked to tell a fellow retreatant why we were there. I said I was researching silence for a book, my fellow retreatant announced she was struggling with a bereavement, and then we were told to be silent. That bombshell, that glimpse into her heart, was left hanging in the air for three days. The only two words I said in the next twelve hours were "compost bin?" as I searched the kitchen during my work hour. The only other talking I did over the following three days was during two brief question-and-answer sessions with the teachers.

On the first evening, I was struck by how odd it was going around the house without talking. There were about fifty of us, so I was constantly passing people in the corridors, or waiting in line for food and the bathrooms, but no words were exchanged. I did more smiling at strangers in one day than I normally do in a whole month, but eye contact alone seemed odd and awkward.

Sitting down to our simple Buddhist meal of split-pea soup and whole-wheat bread (maybe not the best choices for maintaining bodily silence), I found myself directly opposite a woman in her midforties. I did not know where to look. We were close enough to be invading each other's space, but being unable to say hello made the proximity seem exceptionally intrusive. It was weird not being able to engage in small talk. The retreat teachers encouraged us to find comforting community in the shared experience and something supportive in the silence. For me, however, it was a struggle, and I had a strong feeling of cold isolation.

The room where we meditated was the size of a small church, and we sat or knelt in regimented rows on mats. Everyone had a personalized nest shaped from cushions, blankets, and tiny wooden stools to make the sitting more bearable. At the front the teacher sat mostly in silence, occasionally giving instruction. As I settled down for the first sitting, I realized that not only was I physically ill prepared, but I also did not know how to meditate. "How do you know you have a body, right now . . . how is your breath?" the teacher queried precisely and slowly. I learned self-hypnosis twenty years ago, and around the same time I had practiced the Alexander technique to improve my posture, so I did my best to meditate by mingling these techniques with hints picked up from the teacher.

The teacher asked how we knew we were "inhabiting a body." Apart from discomfort and breathing, the awareness came from the sounds around me. This was hardly a silent retreat! Above the meditation hall was a large rookery, and the loud squawks and squeals as the rooks fed their chicks rang out across the hall, interspersed with the soft, warbling tones of blackbirds and the cooing of wood pigeons. Less poetic was the gurgling pipework—from stomachs and radiators—and coughs as people cleared their throats. As I was to learn over the next few days, part of meditation is accepting these sounds and incorporating them into the practice.

On the way to the retreat, I had read some scientific papers about how brain networks are altered by mindfulness techniques, and these accounts helpfully described the stages of focused-attention meditation, which I then copied.[35] You start with a focus—say, your breath passing through the nostrils. Inevitably, your mind wanders. When you become aware of being distracted, you need to shift attention back to the focus. Different brain regions are involved in each of these stages. In an experiment carried out by Wendy Hasenkamp, subjects meditated for twenty minutes in an fMRI scanner to measure brain activity. The subjects were asked to press a button whenever they realized their mind had wandered, before returning to their breath focus. Experienced meditators had more connectivity across networks of brain regions that might be used for maintaining attention and avoiding distraction.[36] This increased connectivity might have been in place before the people started their years of practice, in which case it might be evidence that they were well suited for meditation. Alternatively, it could be evidence that meditation changes neural structures. Attention is not just important to meditation; it plays an important role in cognitive processing. Many aspects of it—alerting, disengaging, reorienting, maintaining attention—are also useful elsewhere in life.

Having survived the first few meditations, I grabbed a quick barley-and-chicory coffee substitute (I bet your mouth is watering!) and retired to the lounge. It was like a dreadful retirement home when the television is broken. Chairs were pushed against the walls and we all sat there staring at our cups, at the walls, or through the windows into the darkening green hills outside. I decided to go to bed early. I was sharing a room with two strangers, and I could not even say good night. It was like a scene from a 1970s sitcom where a marriage had gone wrong—or in this case, a same-sex three-way civil partnership. We padded around the bedroom not looking at each other or talking, passing like ships in the night.

For some there is joy in this communal silence—the freedom from having to put on an act. The silence creates anonymity, since you do not know people's names, where they come from, what jobs they do, and so on. Taking a break from being mindful at breakfast, I glanced around and tried to guess who everyone was, but the loose, formless meditation clothes offered few hints. A young man sat in a fleece sarong and woolly hat, a thirty-something woman had a tie-dyed top and leggings, and an older man sported a goatee and looked as though he played in a traditional jazz group. It was like living in a whole-food shop.

Half of the sessions were walking meditations, which were better outside, even when it was drizzling and cold. The idea was, while walking, to notice how the feet struck the ground and how the lower legs moved and were braced for each step. A cyclist passed on the lane just outside the grounds and stared as the retreatants walked purposefully and incredibly slowly in random directions. Adding to the chorus of birdsong, a malevolent hum came from a tree in full blossom as insects buzzed about pollinating, and wing-beat noise was just above my head.

Maintaining silence between meditation sessions encourages continuous mindfulness. At the time, I was so busy being mindful that it was hard to judge what effect all this silence might be having on me. Only after leaving the retreat did I notice the effects. The sandwich I bought at the railway station while traveling home tasted exceptionally strong.

The idea that meditation can change basic perception is gaining traction in the scientific literature, although not many results are available yet, and there is nothing on taste or sound. Katherine MacLean and collaborators have looked at one aspect of vision, testing people on a three-month Buddhist samatha meditation retreat in a remote mountain setting in Colorado. They had retreatants look at different-sized white lines on a black screen and categorize each line as long or short. By the end of the retreat, compared to a control group, retreatants had improved their ability to discriminate different line lengths, and five months later they still demonstrated improved acuity.[37]

My family laughed at me when I arrived home from the retreat because I spoke in an uncharacteristically soft voice and walked about at a snail's pace. Immediately after the retreat, I felt it had been an interesting experience but not one to repeat. In the weeks and months that followed, however, I had a lingering desire to spend another weekend in that noisy silence, to take the time to rediscover the peaceful state in which I had arrived home.

8

Placing Sound

I f I were to ask you about iconic images of London, Paris, or New York, you might name famous landmarks like the Houses of Parliament, the Eiffel Tower, or the Statue of Liberty. But what about iconic sounds? How readily can you name *soundmarks*, those keynote sounds that define a place and make it special? Sound-marks can be as varied as landmarks: in Vancouver, Canada, the Gastown steam clock marks time not with bells but with whistles; on the Orontes River in Hama, Syria, ancient waterwheels called norias let out loud groans as they slowly rotate, and my travels in the southwestern US were punctuated by the dissonant hoots of Amtrak trains.

For Great Britain, the bongs of Big Ben, the giant bell housed within the clock tower of the Houses of Parliament in London, are a signature sound. In the UK, Big Ben rings in the New Year, has been broadcast at the start of news bulletins for decades, and starts the two-minute silence on Remembrance Day (Britain's Veterans Day). What makes chimes of bells so special? The answer to this question is partially social (bells have played an important cultural role for millennia), but there is also something special about the sound itself. Listen to a bell chime, and what at first seems a simple ring is actually a very complex sound. Why does it start with a clang? Why does the ringing have a dissonant warble? What role does the anticipation of the strike play in our perception of a great bell?

These were the questions going through my head as I approached the Houses of Parliament. Winter sunlight was glinting off the gilded ornamentation around the clock faces on the day of my audience with Big Ben. The outside of this Gothic revival tower is a grand Victorian edifice, a popular establishing shot for film directors to portray "now in London," but the inside is utilitarian. More than 300 stone steps spiral up a narrow staircase to the belfry. The tour took a breather halfway up the tower in a small room that wraps around the staircase. There, our guide, Kate Moss (not the supermodel), told us about the amazing engineering behind the clock's construction in the mid-nineteenth century.

The astronomer royal at the time, Sir George Airy, had set exacting standards. The first stroke of each hour had to be accurate to within one second, and he insisted on being telegraphed twice a day so that he could check the timing. Such precision was greater than that of similar clocks of the day, and difficult to achieve because wind pushing on the 3- and 4-meter-long (10- and 13-foot) copper clock hands can change the speed of rotation. Barrister and gifted amateur clock maker Edmund Denison came up with the solution: the Grimthorpe escapement, which isolates the giant swinging pendulum in the middle of the tower from the vagaries of the weather.

Rested, we climbed more steps before stopping in a tight corridor behind the clock faces. A clunk from the clock mechanism was a signal that the next strike of the bell was two minutes away, so we quickly climbed the final flight of steps up into the belfry. This simple, functional space, with scaffolding and wooden walkways, is completely open to the elements, and a bitingly cold wind whistled through.

The great bell is 2.2 meters (7 feet) high and 2.7 meters (9 feet) in diameter, and it weighs 13.7 tonnes (about 15 tons). Since we were standing only a couple of yards away from the metal, Kate handed out earplugs to protect our hearing. Four bells in the corners of the belfry produce the famous Westminster Chimes before the great bell is struck. Kate instructed us to listen for the third chime bell, which would signal a good time to put in our earplugs. My anticipation and excitement built during the long pause between the Westminster Chimes on the corner bells and the bongs from Big Ben. A large, 200-kilogram (440-pound) hammer drew back slowly before crashing forward and striking the outside of the bell. Even with earplugs in, the sense of power was visceral. The sound resonated the air in my chest like a pumping bass line in a nightclub.

With ten strikes to enjoy, I could examine the quality of the bongs in detail. At first there was a clank of metal on metal, which gradually faded into a sonorous ringing that lasted about twenty seconds. While the initial hammer blow created sound with lots of high frequencies, these rapidly died away, leaving a gentler low-frequency ring, which slowly warbled.

The start of a musical note, the "attack," can be a fleeting moment, but it is incredibly important. As a saxophonist, I spend a lot of time practicing the correct way to start a note cleanly by coordinating the right air pressure from my lungs with precise use of my tongue on the reed. For a violinist, the attack is about the start of the bowing action; just listen to someone learning the violin to hear what getting it wrong sounds like! The attack is one of the main things that makes a sound individual. Big Ben's clang is as much a part of its soundprint as the long warble and the Westminster melody.

Eight months after hearing Big Ben's crisp attack, I heard the complete opposite at a sound artwork. There are only a few examples of permanent sonic art in the world, and three of them are wave organs—in San Francisco in the US, Zadar in Croatia, and Blackpool in England. Blackpool is the archetypal British seaside resort, with its fish-and-chip shops, amusement arcades, and miles of sandy beach. It divides opinions: some see it as a mecca of low-brow fun; others, as the epitome of tackiness.

It was a typical English summer's day when I visited: I had to wear a waterproof jacket to protect me from the cold wind rushing off the Irish Sea, and there were only intermittent glimpses of the sun. The organ stands behind a parking lot next to the promenade, with the UK's tallest and fastest roller coaster generating distant screams from across the road. A narrow, rusting sculpture, 15 meters (50 feet) tall, shaped like a fern in spring beginning to unfurl, forms the most visible section of the wave organ (Figure 8.1). The sculpture is a popular spot to stop and light up a ciga-

Figure 8.1 The high-tide organ in Blackpool.

rette because it offers shelter from the wind. When I first arrived, the organ was only letting out the occasional groan. "Sounds like a moaning cow," remarked a young woman as she walked past.

High up in the rusty fern, I could see a few organ pipes, just like the ones you might see in a church. To get a better sense of what was going on, I climbed onto the tall seawall. Trailing down the concrete sea defenses was a set of black, plastic pipes that disappeared into the water. As the sea swells up, it compresses air in the plastic pipes, pushing the air up to the organ in the fern. Just as happens in a church organ, the air forced into the bottom of an organ pipe by the sea waves meets a constriction just below a rectangular slit in the side of the tube. The fast jet created there causes the air in the main part of the pipe to resonate, creating the tone.

For any organ, the air needs to get up to speed quickly so that the pipe can start speaking cleanly. In this case, however, with the artwork being driven by water waves that flood and ebb irregularly, the notes often start tentatively and die away erratically—hence the groaning and moaning.

The Blackpool organ is an aural rendering of the tidal conditions, a "musical manifestation of the sea," according to the plaque on the side. So I waited around to see what would happen as the tide receded. After about half an hour, the water's edge had dropped, and the movement of the water was more vigorous in the plastic tubing. The organ pipes that are tuned to higher-frequency notes began to play. The overall effect was now like a lazy orchestra of train whistles, or a slow-action replay of a nightmare recorder lesson.

In another half hour the water's edge only just covered the plastic pipes, and the organ was positively energetic. It produced random notes rapidly in an almost rhythmic pattern. The organ pipes were tuned so that the notes could blend together, but overall the sound reminded me of the simplistic computer-generated music I had made as a teenager—not something that many people would listen to for a long time, because the pattern of notes was too random. As discussed in Chapter 7, music works by subverting our expectations. Our brains enjoy hearing the unexpected, but only within reasonable limits. Listeners need an internal schema in their head for how things should be, that can then be subverted as the music progresses.[1] The random notes from the tide organ were just too unpredictable. As one of the artists who designed the work, Liam Curtin, said, "It will be an ambient musical effect and not a popular melody." Furthermore, the organ never reprises its tune. Says Curtin, "On stormy days the performance is wild and frenzied and on calm days the sound is softer."[2] After a while the organ fell silent, as the sea dropped below the level of the plastic pipes on the beach.

The attack of a sound helps us identify the source, whether that is a groaning wave organ, a conventional musical instrument, or Big Ben. Listen to trumpets, bowed violins, and oboes with the early parts of the notes artificially removed and they sound very similar, something like an early synthesizer from the 1980s. The initial scrape of the bow on the string, or the puff of air that parts the reed of the oboe, gives vital cues as to which instrument is playing. In the case of Big Ben, the rapid change in frequencies after the hammer has struck as it settles into a bong is the first cue that we're listening to a bell.

Many large bells warble. Six months after hearing Big Ben, I heard a distinct warbling effect when the Great Stalacpipe Organ played its hymn in Luray Caverns (see Chapter 2). The complicated shape of a stalactite close to me created two notes of almost the same frequency. The resulting tremor, known as *beats*, is caused by a simple addition of the sound waves, as illustrated in Figure 8.2. One note I analyzed had sound at 165 hertz and 174 hertz. These are close enough in frequency that a note at the average of 169 hertz

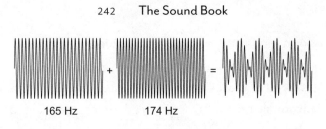

165 Hz   174 Hz

Figure 8.2 Two notes adding together to cause beats.

is heard, with a loudness that rapidly changes at a rate given by the difference in the frequencies (9 hertz). A subtle warbling was added to the stalactite ring, adding a touch of sci-fi spaceship to the musical note.

Guitar players can use beats to help tune their instruments. They press the low string at the fifth fret and leave the next string open, plucking both notes simultaneously. If the two notes are slightly out of tune (that is, not at the same frequency), they will produce a warbling, which is caused by beats. Correctly adjusting the tension in one of the strings brings the two sounds closer in frequency. With a difference of about 1 hertz, the beats are slow enough to impersonate by saying "wowowowowow." The beats get slower and slower as the notes become nearer in frequency, until they disappear altogether when the strings are in tune.

For a bell, symmetry, or rather lack of it, causes the warble. If the bell does not form a perfect circle, it rings with two similar frequencies that beat together. When casting a new church bell, a Western foundry would normally want to avoid such a tremor. But in Korea, the effect is seen as being an important part of the sound quality. The King Seongdeok Divine Bell, cast in AD 771, is better known as the "Emille." "Emille" means the sound of a crying child, and legend has it that the maker had to sacrifice his daughter to get the bell to ring.[3] Big Ben beats distinctly because it creates two frequencies stemming from imperfections, with one flaw clearly visible. A large crack opened up on one side shortly after the bell was first installed. George Airy instructed that the bell be turned so that the crack was away from striking point, that a lighter hammer be used, and that neat square cuts be made at the ends of the crack so that it would grow no further.

The slowly decaying ring of Big Ben produces a less dulcet tone than long notes played on wind, string, and brass musical instruments. A note is actually made up of a combination of sounds at different frequencies. There is a *fundamental* plus additional *harmonics* (overtones), which color the sound and change the timbre.[4] Low notes on a clarinet sound "woody" and quite unlike a saxophone, even though both the clarinet and the saxophone are wind instruments driven by single reeds. The clarinet is a cylindrical tube, so it produces a different pattern of harmonics than does a saxophone with its conical bore. Comparing the harmonics of musical instruments and bells can help explain the differences in the sounds.

Figure 8.3 presents an analysis of a note played on my soprano saxophone, showing the fundamental peak to the left and a whole

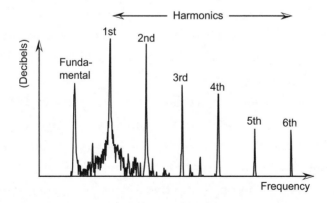

Figure 8.3 A single saxophone note. (Sometimes the fundamental is called the first harmonic, in which case the subsequent peaks would be labeled second, third, fourth, . . . harmonics.)

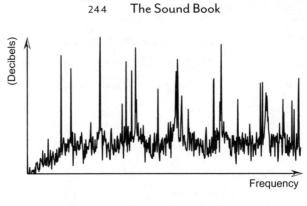

Figure 8.4 The ring from Big Ben.

set of spikes to the right that represent the harmonics neatly and regularly spaced in frequency. In contrast, the analysis of Big Ben's ringing shown in Figure 8.4 reveals a forest of irregularly spaced spikes. The interplay between these harmonics is one reason a bell has a dissonant, metallic sound.

When two notes being played together appear to be fighting each other and clash, this is dissonance. At the core of Western music is the switching between tension-filled *dissonance* and harmonious *consonance*. A good example is the sung "amen" at the end of hymns, where the notes sung under the "a" feel unfinished, and the sound resolves only when the notes switch to those accompanying "men." This feeling of tension being resolved is something we tend to enjoy.

When two notes are played simultaneously, the sounds merge as they enter the ear canal. How we respond to the combined sound is partly dictated by how the harmonics align in frequency. For a simple interval like a perfect fifth (see Figure 8.5), where the notes sound pleasantly consonant together, the two sets of harmonic frequencies are nicely spaced.

For a dissonant interval like a major seventh, however, the pattern of harmonics from the two notes is uneven (see Figure 8.6), with some peaks being close together. In the inner ear, where vibrations are turned into electrical impulses, sounds of similar frequency are actually analyzed together in ranges called critical bands. If two harmonics end up in the same critical band, but not at exactly the same frequency, then a rough, dissonant sound results.

Dissonance and consonance are also exploited in sound art. *Harmonic Fields* is an artwork by French composer Pierre Sau-

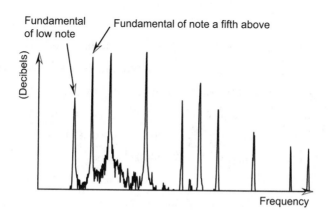

Figure 8.5 Two combined saxophone notes that sound consonant.

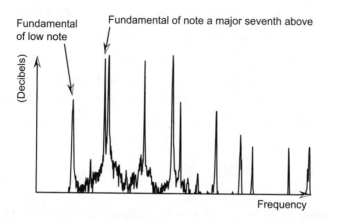

Figure 8.6 Two combined saxophone notes that sound dissonant.

vageot, which I visited six months before my trip to Big Ben. As I approached by bus, I could see a forest of wind-driven musical instruments atop Birkrigg Common, a hill near Ulverston in the English Lake District. In medieval times the raised location would have been a good spot for a castle, but now it is a prime location for catching the prevailing westerly winds. Leaving the bus behind, I walked up the hill with some trepidation. The air seemed very still, and I feared the instruments would be silent. But as I approached a tall scaffold with baubles hanging from drooping metal branches, I was relieved to hear it hum.

Harmonic Fields is a huge artwork with hundreds of different musical instruments. It has few visual charms—just industrial-looking wires, orbs, and scaffolding set out here and there in a confusing pattern. The artist requests that visitors do not take pictures but rather concentrate on the sounds. I slalomed through lines of vertical bamboo poles, which whistled like a panpipe ensemble on opiates. They made sounds like a flute; as jets of wind hit the edges of slits in the wood, the air column inside the bamboo resonated. I walked the length of a wire that looked like a zip line, pausing to poke my head into a drum, which was attached to the middle. The drum amplified the vibration of the wire, producing a tone just above middle C, a note in the center of a guitar's range. But the hum was not constant; it waxed and waned, sounding like a wet finger being run around the rim of a large wineglass.

My favorite piece was very simple and unassuming. Plastic strips were strung between tripods like clotheslines. When I first walked over to it, I kept looking up for the helicopter overhead that was ruining my recording, until I realized the "whop-whop-whop" was actually coming from the strips themselves, which were acting as a giant Aeolian harp.

When wind passes over a wire, the air above and below has to speed up to get around. A fast-moving stream of air flows away

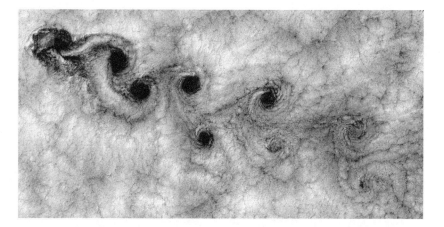

Figure 8.7   Satellite image of clouds showing airflow around Alejandro Selkirk Island, Chile. (The island is in the top left of the image. The whorls behind it are oscillating wakes, a visual demonstration of turbulence.)

behind the wire, moving into the space just behind the wire, switching between two states: first the airstream from above fills the space, and then the stream from below. This alternating airstream causes the wire to vibrate back and forth and create a tone.[5] The same phenomenon happens on a vast scale as air flows past islands, as can be seen from satellite pictures of clouds like the one in Figure 8.7. For the Aeolian harp, both the frequency and the loudness of the sound change as the wind speeds up and slows down, so the tone is forever altering.

I once presented a radio program titled "Green Ears" about sound in the garden, and all the experts I interviewed hated wind chimes. For those gardeners, some parts of *Harmonic Fields* would have been like wind chime hell, where a forest of glockenspiels rang with the maniacal strikes of mallets driven incessantly by turbines. Pierre Sauvageot sees the whole work as a musical composition, "a symphonic march for 1,000 [A]eolian instruments and moving audience."[6] Hence, the wind instruments are carefully tuned to play particular notes, which combine to produce in some places sweet harmony and in others, a malevolent dissonance like a swarm of insects.

Because consonance and dissonance underpin music, they have also played a central role in the debate over why humans have evolved to like music. Thomas Fritz of the Max Planck Institute for Human Cognitive and Brain Sciences in Leipzig, Germany, wanted to see how people who have not been exposed to Western music responded to consonance and dissonance, so he traveled to Cameroon in Africa to study the Mafa people. The Mafa are located in the extreme north of the Mandara mountain range. The most remote settlements have no electricity supply and are culturally isolated by pervasive illnesses such as malaria. The Mafa make ritual sounds, and Thomas once played an example for me. It sounded like a dissonant chorus of old car horns, but actually the noise came from vigorously blown flutes. Thomas compared the responses of the native Africans to Westerners, playing both groups a variety of music styles from rock and roll to the Mafa's ritual sounds, including versions of all the pieces that had been electronically processed to make them sound continuously dissonant. Both groups showed a preference for the less dissonant original pieces over the manipulated versions.

From a Western viewpoint, the story seems simple. We find dissonant sounds unpleasant because the bias is "hardwired" into the brain, and this preference underpins music compositions. But in recent times, some scientists have pointed out that many cultures actually embrace dissonance. I once interviewed Dessislava Stefanova, leader of the London Bulgarian Choir, for a BBC radio program. She and a colleague demonstrated the technique of "ringing like bells." They sang two notes that set up the strongest dissonance I have ever heard. Analysis of the sound shows that the notes occupy the same critical band of the inner ear, with a frequency spacing to maximize dissonance. But instead of resolving this dissonance to consonance, the singers just left it hanging in the air. They were enjoying the dissonance and felt no need to resolve it.

Weighing the current evidence, I believe humans initially find consonance pleasant and dissonance unpleasant, but this innate preference can be changed by music that we hear during our lives, starting with what we hear in the womb during the third trimester. This raises the question of why we find consonance pleasant to start with. What evolutionary drivers might cause this preference? Although news articles frequently ascribe human characteristics to evolution, peering back through the ages with any scientific certainty is usually impossible. But this impediment has not prevented us from speculating. One theory is that the response is a by-product of the auditory system's being trained to understand speech in noisy places.[7] After all, there is a close relationship between speaking and singing, with the vowel sounds being virtually sung during talking. This theory could also fit with the view of experimental psychologist Steven Pinker. He famously described music as "auditory cheesecake," something that is pleasurable but has no adaptive function, coming as a by-product of other evolutionary pressures, like learning language.

I find it hard to believe that music serves no evolutionary purpose. Charles Darwin thought music was a sexual display, an equivalent of the elaborate courtship calls of animals such as the superb lyrebird of Australia. The male lyrebird builds a stage in the rainforest from which he pours out the most remarkable song that is an amalgamation of things he has heard. He impersonates the calls of about twenty other species, including whipbirds and kookaburras, and even mimics the sounds of camera shutters, car alarms, and foresters' chainsaws. But while music is often about love and sex, it goes way beyond that to an abstract art dissociated from reproduction. When I went to the performance of John Cage's *4'33"*, I got a strong sense of a collective endeavor with the rest of the audience. Robin Dunbar at the University of Oxford argues that music making plays an important role in social bonding, and the ability of humans to work collaboratively is one of our reasons for evolutionary success.[8] Music also plays an important role in developing bonds between parents and babies, from soothing lullabies to the exaggerated intonation in motherese (baby talk), which helps babies learn to speak a language.

Whatever the origins of our love of music, it is known to have a strong effect on us. It activates more of the brain than any other known stimulus. Music we like excites the reward centers that release the chemical messenger dopamine. A similar response is seen with other pleasurable activities, like sex, eating, and drugs. Did my brain respond in this way to the bongs of Big Ben? Neuroscientists have yet to study in detail our response to bell sounds and other soundmarks. But given our emotional relationship to natural sounds and other familiar noises, I would expect there to be a neurochemical relationship between soundmarks and pleasure, even for the mildly dissonant warble of Big Ben.

Important and powerful institutions, such as town halls, churches, and monasteries, use bells to signal time, to herald the start of religious services, and to mark important historical events. Bells might be rung to warn communities of danger, call men to arms, celebrate military victories, or honor the passage of lives through christenings, marriages, and funerals. Alain Corbin, who studied the role of bells in rural France in the nineteenth century, makes persuasive arguments that the sonic footprint demarcated a local community's territory both socially and administratively. Bells were used to mark the end of the workday, so townsfolk had to stay within earshot.[9]

While church bells are heard worldwide, they are usually just chimed, where the mouth of the bell hangs downward and the clapper bounces back and forth inside. Bell ringers use a different set-up to create the quintessentially English sound known as *change ringing*, which originated in the sixteenth century and can be heard every weekend at churches throughout the country. Change ringing produces peals of sound in rhythmic patterns from a set of bells, like an ancestor of minimalist compositions by Steve Reich or Philip Glass.

I have always wanted to know more about change ringing, so one autumn afternoon a couple of months before my trip to Big Ben, I headed down to a church close to my home. St. James' is a Gothic-style village church serving one of Manchester's leafier suburbs. Ignoring the jam-and-cake stall outside the front door, and the exhibition on weddings in the nave, I climbed up an extremely tight spiral staircase, ducked under a very low doorway, and entered the ringing room. Hanging through holes in the ceiling were six thick ropes, each with its own woollen grip, or "sally." Paul, one of the regular bell ringers, eruditely explained the practice of change ringing. Another of the regular ringers, an enthusiastic fellow named John, had a tabletop model that he demonstrated to me. I could also see what was happening above in the belfry via a webcam.

Each rope connects to a bronze bell in the belfry through a hole in the ceiling. Although the six bells in St. James' form the first notes of a major musical scale, melodies are not the goal. A team of half a dozen campanologists pull the ropes and ring the bells in different orders following a mathematical pattern. Opposite Paul was a whiteboard covered in bewildering grids of colored numbers connected by lines, showing examples of the order in which the bells were to be rung. As Paul explained, ringing the bells is a disciplined and regimented process with, "ears open, eyes wide on the white board."[10]

John and Paul can precisely control when a bell tolls because it is mounted on a large wheel that allows it to rotate a full circle. Before ringing, John used his rope to turn the bell upside down so that the opening pointed upward. There it waited until he pulled the rope again, tipping the bell over; it then turned a full circle before finishing pointing upward again. Another tug on the rope then turned the bell in the other direction for a full rotation in reverse. The bells are incredibly heavy, and John explained to me that I had to work with it rather than fight against it. Since I had never done this before, I got to do half the job, the back stroke. John would pull the bell around once, and then I would pull it back the other way. The rope dangled down between my legs and I held the sally like a cricket bat. When John pulled the rope and created the forward rotation, the inertia of the bell pulled my arms above my head; I tried to pull it back but completely mistimed my effort and struggled to heave the bell around. After a few more attempts, I hit a rhythm. When the bell just starts tipping in the right direction, a long, gentle pull will bring it around full circle.

Wanting to know more about people's responses to peals of bells, I turned to sound artist Peter Cusack, who started exploring the general public's response to sound in London about a decade ago. His investigative technique is deceptively simple. He just asks, "What is your favorite sound of London, and why?" As well as identifying stuff for Peter to record, the question reveals personal stories about the sounds. The Favourite Sound Project has been expanded and run by others, to involve cities around the world, including Beijing, Berlin, and Chicago.

In London, respondents to Peter's question often mentioned Big Ben, but not always the actual sound of the bell. People remembered the time between the strikes—the moment of anticipation before the next bong, when the auditory cortex increases activity as it directs attention waiting for the sound—something I experienced intensely in the belfry. The sound of Big Ben on the street is quite different because the impact is diminished by traffic noise. When the great bell first tolled about 150 years ago, Londoners could hear it from farther away than is possible today. Iconic sounds are much more localized nowadays because of the blanket of noise in our cities.

Cockneys are the working-class people from the East End of London famous for their rhyming slang, saying "apples and pears" to mean "stairs," "plates of meat" for "feet," and "trouble and strife" for "wife." To be a true Cockney you have to be born within the sound of the bells of St. Mary-le-Bow Church. But an acoustic study implies that Cockneys could soon be "brown bread" (dead), because the area over which Bow Bells can now be heard is so small that it contains no maternity hospitals.[11] A hundred and fifty years ago, London was as quiet as the countryside is nowadays, with a level thought to be 20–25 decibels in the evening, and it is estimated that the bells could have been heard up to 8 kilometers (5 miles) away. Nowadays, the noise from roads, aircraft, and air conditioners mean that the London sound level is usually about 55 decibels, and the bell tolls travel audibly across a mile at best.

Six months before I heard Big Ben, I found myself only 500 meters (550 yards) from St. Mary-le-Bow, listening to a sound sculpture called *Organ of Corti* (Figure 8.8). Designed by Francis Crow and David Prior, it aimed to sculpt and recycle environmental noise such as the traffic sounds that mask the bells of London. *Organ of Corti* was made from ninety-five transparent, vertical acrylic cylinders, each about 20 centimeters (8 inches) in diameter and rising to a height of 4 meters (13 feet). Named after the part of the inner ear within the cochlea that responds to sound, the sculpture reminded me of a giant child's plastic toy, with the forest of translucent cylinders distorting the bodies of people walking through.

The sculpture's design exploits a scientific discipline that was sparked by another artwork. Installed in Madrid in 1977, Euse-

Figure 8.8 *Organ of Corti.*

bio Sempere's *Órgano* is a large, circular forest of vertical steel cylinders. Not until the 1990s did measurements by Francisco Meseguer, of the Institute of Materials Science in Madrid, and his colleagues reveal that this minimalist sculpture shapes sound. Meseguer normally works on photonic crystals, tiny structures that alter light. Shine white light onto one of these crystals, and some colors become trapped inside and do not pass through to the other side. If you pick up a peacock tail feather and twist it around in your fingers, you will notice how the color changes. Microscopic periodic structures create this iridescence. In nature, the most striking colors on butterfly wings, squid bodies, and hummingbird feathers are made by photonic crystals and not pigments.

A conversation with acoustics expert Jaime Llinares made Meseguer realize that if photonic structures were scaled up, they would make a sonic crystal, stopping sound at particular frequencies from passing through. In 2011, I showed that sonic crystals also reflect some frequencies intensely, mimicking the iridescence of butterfly wings (unfortunately making an unpleasant sound).[12] *Órgano* was just the right size for Meseguer and Llinares to test out their idea, being about 4 meters (13 feet) across, with the cylinders regularly spaced about 10 centimeters (4 inches) apart.[13]

Meseguer placed a loudspeaker on one side to produce noise. A microphone on the other side confirmed their hypothesis: there were band gaps, frequencies of sound that did not go through the array of cylinders. The effect is caused by interference, a phenomenon first explained by British physicist Thomas Young in 1807. Young was a child prodigy who could speak fourteen languages before he was nineteen and first trained as a physician. His classic double-slit experiment, which is still used in schools to teach physics, is illustrated in Figure 8.9. When a monochrome light is shone through two slits, a pattern of light and dark appears on a screen. In some places waves from the two slits arrive with their peaks

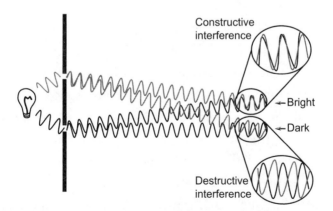

Figure 8.9 Thomas Young's double-slit experiment.

and troughs aligned and there is constructive interference, giving a bright patch. Sometimes the two waves are misaligned and they cancel each other out, leading to a dark patch caused by destructive interference.

The same effect can be demonstrated with sound if you use a loudspeaker and a screen with the slits farther apart. You can go further and add more slits in the screen, and even many perforated screens spaced out one in front of the other. If you then remove the screens and place a cylinder where each slit was, you have made yourself a sonic crystal like Eusebio Sempere's sculpture. As in the double-slit experiment, what is heard passing through the forest of cylinders is determined by constructive and destructive interference, with sound of some frequencies getting trapped inside, rattling back and forth between cylinders and failing to emerge.

As soon as people discovered that sonic crystals block sound, experimenters began testing their use as noise barriers. But crystals attenuate only a few select frequencies. Consequently, solid barriers made of wood and concrete are nearly always more effective at stopping broadband noise. A colleague of mine at Salford, Olga Umnova, has been experimenting with an acoustic black hole that absorbs a wider range of frequencies. The black hole is made by gradually decreasing the diameter of the cylinders at the edge of an array of crystals. The result is an outer shell that guides the sound to the core, where it can be removed using conventional absorbants.[14] Sonic crystals have also caught the media's attention as a way of making the sound equivalent of Harry Potter's cloak of invisibility. Normally, we can sense the presence of an object by hearing the sound reflecting from it. The "cloak of inaudibility" surrounds the object, gracefully bending acoustic waves around it so that it cannot be heard. Unfortunately, sonic crystals are often too big to be practical, because sound waves are quite large, especially compared to light.

Organ of Corti had some of its acrylic cylinders missing to form a sinuous path through the middle of the sculpture. Being a regular array of cylinders, the artwork should have amplified certain sound frequencies and attenuated others. But I had chosen a bad day for listening in London. Workers were digging up the road using jack-hammers, and as I tried to listen for subtle changes in the noise, the power tools stopped and started randomly, making it impossible to tell what the sonic crystal was doing.

Later that summer the sculpture was placed near a weir on the River Severn in England. The more constant noise from the waterfall made it easier to hear the sound being sculpted. Francis Crow told me that the effects were most obvious as you exited the sculpture. Walking into the structure, the forest of cylinders subtly removed selected frequencies from the noise, but such absences could be hard to hear. Walk out of the artwork and these removed frequencies reappeared and could be heard. This makes sense, because our ears are designed as an early warning system to listen for new sounds, not subtle absences.

Francis explained to me that some of the motivation behind the work was to change how people listen: "It's hearing what's already there but framed through the structure."[15] It was the sonic equivalent of *Skyspace*, a series of works by James Turrell. These are large rooms where visitors view the sky through an aperture in the ceiling, framing light and space. *Organ of Corti* was framing how we listen to sound. Appreciating the work takes time; near the weir, people tended to linger and enter a meditative state. One visitor took this idea to an extreme and spent over half an hour inside the sculpture, commenting, "I made my own symphony." Another described the subtle ebb and flow inside as "disorienting." It is difficult to experience sound art quickly. A brief glance at a piece of visual public art is likely to yield more than a brief encounter with a sonic artwork.

Many minimalist sculptures distort sounds. Several of Anish Kapoor's artworks are large, concave light mirrors. I went to see *Her Blood* (1998) at the Manchester Art Gallery in England. The work is made from three huge, concave dishes 3.5 meters (11 feet) in diameter propped vertically against the gallery walls. Two were highly polished mirrors; the third was stained deep red. As visitors walked toward a dish, the visual reflection of their figure was distorted. Some distance away, they could see themselves smeared across the lower half of the dish; as they walked closer, suddenly their image formed rings going completely around the dish. At this position they were at the focal point for both light and sound. Attendants had noticed the way the reflection distorted voices, and they were encouraging visitors to speak at the dishes.

The concave dishes of *Her Blood* did not behave much differently than the radome in Teufelsberg. In contrast, Richard Serra's giant works at the Guggenheim Museum in Bilbao, Spain, produced an astonishing diversity of sounds. It was like having a giant sound-effects unit to play with. *The Matter of Time* (2005) is an installation of seven giant sculptures made from spirals, serpentine ribbons, and sinuous curves of rusting steel that rise many yards overhead. The brown metal walls rise at angles, forming narrow passages that twist and turn, sometimes closing over in an inverted *V*, disorienting a sense of space and balance as you walk through. It is like being in a giant steel maze; around the corners I expected to bump into Alice from Wonderland.

When I visited, the gallery echoed with the hubbub of chattering schoolchildren. As I entered some of the works, I could hear the effects of going into a partial enclosure: the ambient noise quieted, and my ears sensed the quick reflections from the steel walls. The sound was being reframed as happened with *Organ of Corti*.

I was fortunate to have a press pass, which meant I could take out my digital recorder, wait for a moment when I would not disturb others, and clap my hands to reveal the acoustic. In the middle of the giant spirals were large, round arenas about 8 meters (30 feet) across. In these spaces the reflections focused into the middle. This focusing created Gatling gun echoes, as the claps whizzed past me every twenty milliseconds. In some places, stamping my feet made repeated twangs like vibrations propagating up and down a very long spring. As I mentioned in Chapter 5, some were very good whispering galleries, carrying my voice efficiently from one end to the other, with the sound hugging the walls of steel.

Best of all was *Snake*, a work made from three long, tall, twisting metal sheets forming two narrow corridors about 30 meters (100 feet) long. The passageways were only about a meter wide, and resonances across the narrow width colored my voice. When I stood in just the right place, with a flat bit of the ceiling high above the sculpture, the sound would ricochet back and forth between the ceiling and the floor. Sound would also go along the narrow channel and be reflected from other sculptures at the end, before being returned as a diffuse echo. Stamping my feet on the floor was very satisfying because in just the right place I could impersonate a rifle shot. I was not the only person enjoying the acoustic distortions; others were calling "hola," "echo," and "boo" as they walked around.

When Peter Cusack asked Londoners about their favorite sounds, often the answers were about mundane and everyday sounds. The favorite-sound question is personal, so the semantic meaning of what is heard usually trumps the raw physical characteristics of the acoustic wave. If you take a moment's break from reading and just listen, what do you hear? I can hear voices from the office next door, rain falling on the pavement outside, and footsteps in the corridor. Did you also come up with a list of what made the sounds? I thought "voices, rain, and footsteps," not "mumbling, pitter-patter, and tap tap." We describe what we hear mostly in terms of sources and metaphorical meanings, not their intrinsic sound.

But sometimes the physical characteristics of what we hear do matter. Something loud like an explosion triggers a rapid fight-or-flight response. A plane flying overhead might not be as loud as an explosion, but the sheer volume of the noise might still drown out a conversation. A melody is just a sequence of abstract tones, yet it can tap deeply into our emotions, evoking joy, sorrow, and love. But mostly, for everyday sounds, it is the cause of what we hear that matters. Where possible, the brain identifies the source, and our response is then colored by how we feel about what caused the sound. If you hear a bus in a city square, your response is likely to be strongly influenced by whether you want to get on it, or by your attitude toward public transport: are buses a waste of taxpayers' money clogging up the road or a public good easing pollution and congestion?

This is why Peter's question about sonic favorites reveals sounds that, at face value, have little aesthetic appeal: the announcement "Mind the gap!" in the London Underground, the siren of a New York Police Department squad car, or Turkish market stall holders shouting out offers on Turmstrasse in Berlin.

I am struck by the similarity between the favorite-sound answers and what correspondents told Andrew Whitehouse about birdsong. Many of the stories about city sounds and birdsong are not about the awe-inspiring, surprising, or most beautiful. They are not the sonic equivalent of the Taj Mahal, the Golden Gate Bridge, or the Grand Canyon; instead they are about the sounds that remind us of special places and times, or the sounds likely to be heard in our everyday lives. People often mentioned transportation sounds. After all, getting around a city is such an important part of living and working there. Imagine if the study had asked for favorite images of London; respondents would probably have listed visually appealing or striking attractions such as St. Paul's Cathedral, the London Eye, or Tower Bridge. The ringing of Big Ben is an exception, a sound that has both aesthetic beauty and deep historical, personal, and social meaning for the British.

Peter has been running the Favourite Sound Project for over a decade, so some of the original sounds have disappeared. When trains used to arrive in London, there was a staccato cascade of door slams as passengers disembarked. This sound disappeared as the old rolling stock was decommissioned. What replaces these distinctive sounds is similar across the world because of globalization of technologies and products. Regretfully, cities are becoming more sonically similar and less individual, mimicking the visual homogenization of main street.

Andrew Whitehouse found that birdsong heightened some emigrants' sense of being in an alien country. I got a similar feeling when I traveled to Hong Kong, but with a different sound. My strongest sonic memory was the hubbub of the massed crowds of Filipino women along walkways and crowded into shopping malls. Under the HSBC tower block is a large, covered plaza, which was alive with the high-pitched chatter of females socializing. To Hong Kong residents this is a common sound, as domestic workers gather on Sundays around the city center, lay out their picnic blankets, and socialize with friends. But to me as an outsider, it stood out as something special and unique in Hong Kong.

Under the tower block, the semi-enclosed space amplified the chatter of the women, enhancing the effect. Indeed, soundmarks can be created by the concrete, brick, and stone of the city because they form spaces that alter sound in surprising ways. The Greenwich foot tunnel from Chapter 4 appeared in the list of London's favorite sounds because of the way it distorts voices and footsteps.

When I heard about Italian artist Davide Tidoni, I realized I had found a sonic soul mate, in that he explores the hidden sound effects in urban landscapes. In one of Davide's projects, he bursts balloons and brings spaces to life. A short, loud, impulsive bang is ideal for revealing sonic features. I was fortunate to see him; he had the free time to visit me because he had been forced to take a day off from working in London. Security guards were objecting to his bursting balloons and recording around the Barbican in London.

I decided to take Davide on a walk around the canals of Manchester to sample the acoustics of nooks, crannies, and arches first formed during the Industrial Revolution. After lunch we stopped at a shop to buy some balloons before walking to the towpath of the Rochdale Canal. Fully opened in 1804, this was the first canal to cut across the Pennines, the ridge of mountains that separate eastern and western England in the north. Under a dingy, low, arched bridge I placed my digital recorder on the floor, narrowly missing a discarded condom. Davide then inflated a yellow novelty balloon into the shape of a long, bulbous worm with a couple of tentacles sticking out the top. He waited patiently with a pin poised until the thundering cars driving over the bridge quieted. A series of ricocheting twangs followed the sudden bang as the worm burst and the sound bounced around underneath the arch.[16]

We had spent a long time over lunch discussing the merit of different types of balloons—hence this attempt with the novelty worm. But after the first test, we reverted to using conventional round shapes, preferring the way their shorter, sharper bangs revealed the acoustics more starkly. Davide explained to me that his sonic exploration is about building a relationship with the spaces he is in. He also told me, "What is striking is to see how apparently the same gesture and sound are perceived in different subjective ways depending on the position of the listener and his emotional state."[17]

Davide uses balloon bursts to raise people's awareness of spaces and train their sensitivity to sound. Videos of these walks show people initially being alarmed by the loud bang and wincing—even the person who is plunging the pin into the balloon and therefore knows the explosion is coming. Then people smile, giggle, or stare in disbelief as strange tones and echoes are produced; Davide sees these reactions as a "need to externalize the emotions." On one video a young woman calls out, "Bellissimo!" as she looks up and tries to work out where the sound is coming from. These responses give insight into our hearing. Initially there is a startle response, an unconscious reflex that occurs to prevent injury. People blink to protect their eyes and tense their muscles to brace themselves in case they are about to be physically struck. This reflex is incredibly rapid, going through a very short neural pathway in just 10–150 milliseconds. The slower secondary responses, such as the giggling, occur once the brain has had a chance to properly access the situation and realize that there is no real danger.

Davide has the following wonderful idea of an acoustic gift: "I usually send an invitation to someone that I feel very close to and then we listen together to a place that is particularly meaningful to me." I am not especially close to Davide, but our journey around Manchester was focused on reaching one place, which, I guess, I gave to him as a gift. Built in 1765, Castlefield Wharf is at the end of the Bridgewater Canal. The waterway is commonly cited as the first canal built in England, designed to transport coal to Manchester at the start of the Industrial Revolution. A nineteenth-century railway bridge spans the canal basin, forming a tall, narrow archway where the sound reverberates for a ridiculously long time. We stood under this brick arch clapping and shouting, bewildered by the lingering sound. This narrow space has a reverberation time much longer than a classical concert hall.

Though celebrated by many, the ringing of bells is also a common cause of noise complaints. One case involved the Church of All

Saints in the village of Wrington, England. Built in the late fifteenth century, the church has a square tower housing a peal of ten bells. It had marked every quarter of an hour day and night for a century, but in spring 2012 the bells fell silent after local government officers declared them to be a statutory noise nuisance. Fortunately, a compromise was reached whereby the clock chimes only hourly overnight.[18]

Sound artworks have to be carefully located to prevent similar complaints. I discussed this question with Angus Carlyle, an expert in sonic-arts practice at University of the Arts London. He suggested, "We seem very tolerant of very visible ugliness in the built environment. We also tolerate a mixing and mashing together of visual style in architecture . . . But my hunch is that there would be much less sympathy for an equivalent density of creative sound interventions."[19]

Retired academic Tony Gibbs made a similar point when I phoned him. I called Tony because he has written one of the few academic books on sound art. He argued that the sonic equivalent of large public artworks would have to make a similarly bold acoustic statement—in other words, make lots of noise. As Tony explained, "We as a public, as a culture, don't like massive noises . . . to persuade people that they should regard noise as art is a big ask."[20] For a monumental piece of visual art, such as the 120-meter-high (400-foot) narrow Spire of Dublin, if people don't like its appearance they can simply look away. Screening out sound art would involve reaching for earplugs.

It is a shame that there are relatively few public sound artworks, because some of the most iconic images in the world are sculptures: the Statue of Liberty in New York, the Great Sphinx guarding the pyramids in Egypt, and Christ the Redeemer towering above Rio de Janeiro. In recent decades, governments have turned to public art as a way of bringing communities together, attracting tourism, and aiding or symbolizing regeneration. Striking works have resulted, like Antony Gormley's *Angel of the North*. With a wing-span wider than that of a jumbo jet, this giant, rusting figure towers over Gateshead in England. Is it possible for artists to create the sonic equivalent of this giant work—permanent public sound art that comes to define a place? Something that would rival Big Ben for its iconic sound? Angus Carlyle saw no reason why sound art could not create an "iconic rapport between place and the heard," however, he felt that sound art was too much in its infancy and would have to establish greater prestige before it could attract permanent commissions.[21]

Would public sound art be more accepted if, instead of a loud noise, it produced something more tuneful? Just outside the city of Lancaster in California is a soundmark that creates a rendition of Rossini's "William Tell Overture." Strangely, no electronics are involved. This is a musical road. It creates the melody from wheel vibrations. It is a bit like a rumble strip, the line of ridges off the side of a major road that create a buzzing sound alerting drivers to danger. The musical road uses grooves cut in the asphalt rather than the bumps you get in rumble strips, but the sound is made in a very similar way. The pitch of the notes depends on how fast you're driving and the spacing between the grooves, with close spacing giving a high note, while corrugations spaced farther apart lower the frequency. The road near Lancaster takes the rumble strip one step further by changing the spacing between the grooves in a pattern to create a tune.[22]

The corrugations were first cut for a car ad, possibly inspired by the dozen or so musical roads that exist in Korea and Japan, or the *Asphaltophone*, which was created by Danish artists back in the 1990s. I decided to visit Lancaster to hear the musical road for myself.

It was on a Saturday in June, six months after popping balloons in Manchester, that I turned west off Route 14 onto Avenue G. This is a flat, featureless road a few miles out of the city. After a short distance, a white road sign near a line of trees announced, "Musical Road Presented by City of Lancaster. This Lane ⬂." As my tires rolled over the first few notes, I smiled at this magnificently silly creation. Every time a tire hits a groove in the road, a short sharp vibration travels through the tire, into the suspension, and then into the body of the car. Inside, you hear the sound radiated by vibrations of the cabin interior. The road plays eight bars from the "William Tell Overture." The "March of the Swiss Soldiers" is a frantic gallop, and the road plays the first phrase from the main melody.

I swung the car around and headed back to have another go. Over the next hour, I drove over the grooves half a dozen times trying my microphone in different places. I got the best recording from inside the glove compartment. In this location, the microphone was close to vibrating bits of the interior trim that made the tune louder, while being sealed away from the high-frequency wind noise created as the car cut through the air. Cruise control was useful to set a steady pace, ensuring that the tune did not speed up or slow down as I drove over the road.

As the tire and bodywork of the car are vibrating, they radiate sound outside the car as well as inside. Even with a normal, ungrooved road, tires make noise as they roll over the pavement, and considerable research effort goes into reducing this sound. When I stood on the shoulder, I could clearly hear the melody from passing cars, with the added pleasure of seeing the smiles on the drivers' faces and hearing the pitch of the notes slide about. I stood where the grooves of the first note started, and as a car started the tune and disappeared into the distance, it was as if the first note was sighing as the pitch dropped by three semitones. This is the *Doppler effect*, which is commonly heard with police sirens and fast-moving trains. As the cars drove away from me on the musical road, the sound waves stretched out and the frequency dropped. I

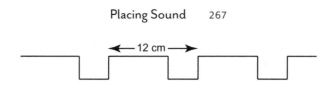

Figure 8.10   Grooves in a musical road.

would have liked to make more recordings, especially when two cars at different speeds were playing the road, because the result was a clash of renditions at unequal pitches. However, it was so windy that it was difficult to stand up, and the gale blowing across the microphone created too much noise for decent recordings.

I have replayed the musical road to lots of people, and many struggle to recognize the tune, even though it is a very famous piece of classical music and was the theme music for *The Lone Ranger*. The problem is the tuning of the notes, with most being at the wrong frequency. As physicist David Simmons-Duffin explains in a humorous blog post, the designers mucked up the tuning because they got the spacing of the grooves wrong.[23] For the lowest note, which starts the tune, the distance between the front edges of adjacent grooves averaged out at about 12 centimeters (4¾ inches), as illustrated in Figure 8.10.[24]

Down the road thirty-six notes later, the tune reaches its high point, with a note that should be an octave above the first. An octave is a doubling in frequency, so the car tire needs to hit the grooves twice as often, which means the distance should be halved to 6 centimeters (2⅜ inches). But on the road the spacing is actually 8 centimeters (3⅛ inches). This means that, instead of an octave, the musical interval is closer to what musicians call a perfect fifth. When you are taught musical intervals, you learn to associate them with particular tunes. So instead of hearing the big leap from the first two notes of "Somewhere over the Rainbow," drivers hear something like the first two notes of the theme music for the film *Chariots of Fire*.

If the interval had been exactly a perfect fifth, then the tune would have simply been disappointingly wrong, but the melody was excruciating to listen to because the sound's frequency was between musical notes, so the road was out of tune.[25] Forcing notes to be at particular intervals and disallowing sounds with frequencies between is vital for most music. In theory, an instrument like a trombone with its sliding valve can produce any frequency within its playing range. But such a free-for-all is unlikely to produce beautiful music. The octave interval is found in nearly all musical cultures.[26] A doubling in frequency is readily processed by the human brain because two notes separated by an octave share the same neurological pathway. This is also true for some other animals. Rhesus monkeys can be trained to recognize simple strong melodies that are transposed by an octave, such as "Happy Birthday to You."[27]

The octave is then subdivided into more notes. In Western music, octaves are divided into twelve smaller intervals called *semitones*, with melodies usually using a subset of these that make up a musical scale. But in Asian music such as gamelan, things are different. The *sléndro* scale divides the octave into five notes and creates a sound similar to playing just the black keys on a piano; the *pélog* scale uses seven uneven intervals. Therefore, the notes found in melodies are not just about some innate processing in the brain; it also comes down to what you have learned by listening. The musical road may have produced frequencies that did not match any notes I know of, but maybe there is a culture where it is perfectly in tune.

While listening curbside was fun as a visitor, imagine if you lived close by. I was actually standing by the second incarnation of the road. The first had been too near houses and, as one resident, Brian Robin, was reported to say, "When you hear it late at night, it will wake you up from a sound sleep. It's awakened my wife three or four times a night."[28] The music produced by the road must have been particularly annoying. Imagine trying to go to bed each night with a garbled "William Tell Overture" going off every few minutes.

Many standards and regulations dealing with noise annoyance have stringent criteria for tonal noise—sounds that have distinct notes. While the brain has a remarkable ability to habituate to hisses and rumbles, tonal sounds are harder to ignore. This is why bells have been used as signals through the centuries: the loud ringing is penetrating and hard to ignore.

On the way down from Big Ben's belfry, I visited the room that houses the clock mechanism, which has another set of wonderful sounds, including the noisy wind governors, which control how fast the heavy clock weights descend inside the tower. The governors are large vanes that rapidly spin against a ratchet mechanism, sounding like a souped-up, old-style football rattle. Machines have revolutionized what we hear, but to assume that all this noise is bad is an oversimplification. As we will see in the next chapter, some technologies are producing the sonic wonders of the future.

9

Future Wonders

Since the Industrial Revolution, our ears have been bombarded with the sounds and noises of technology and engineering. Much of what we hear nowadays—the rumble of a kettle boiling, the ping announcing a new e-mail, or the loud whine of a vacuum cleaner—is artificial and man-made. These sounds are often an accidental by-product of functionality, but increasingly, manufacturers are deliberately manipulating what a customer hears to improve satisfaction and increase sales.

When you view a car in a showroom, the first aural impression you get is not the roar of the engine, but the click and clunk of the driver's door opening and closing as you climb in. About ten years ago, automobile makers realized that the door locks and catches rattled in a tinny way as if they were cheaply constructed. New safety standards had required more hefty side bars in case of accidents, so weight had been removed from other parts of the car to compensate, including from the door catch. Perceptual tests show that people associate well-made products with a bassy sound, maybe because big objects tend to be more powerful and create lower frequencies. To get rid of the tinny sound from the catch, absorbing material was added to the door cavity to attenuate high frequencies, and the locking mechanism was changed so that it closed with a shorter, high-quality clunk.[1]

What about electronic devices that naturally make no sound? Often these are designed to impersonate old mechanical devices. Press the button on a digital camera to take a picture, and you will hear a recording from the shutter of an old film camera. On my smart phone, if I punch in a number on the touch-screen keypad, I hear the sound of an old-fashioned push-button telephone. One thing that could radically change our sound world is the move away from gas engines to alternative fuels. There are fears, however, that hybrid and electronic vehicles are too quiet at low speeds, making it hard for pedestrians to hear them coming.

Companies are experimenting with playing noises from loud-speakers hidden under the hood to alert pedestrians. But what sound should they use? Why, something familiar that immediately makes pedestrians think "vehicle." Nissan has opted for a hum that you could imagine Luke Skywalker hearing from his hovering transporter on the planet Tatooine. In one scientific experiment, however, people preferred the noise of an internal combustion engine over hisses, hums, and whistles.[2] There is a legacy of old sounds surviving from obsolescent technologies. As a correspondent to New Scientist wrote, "Imagine if this concept of familiar sounds had been developed earlier. Would cars all make the sound of horses' hooves instead of the newfangled and confusing drone of an internal combustion engine?"[3]

What is done when there are no old technologies to imitate? Sometimes manufacturers of electronic devices turn to musicians. When composer Brian Eno was asked to write the start-up music for Windows 95, the specification included about 150 adjectives: "The piece of music should be inspirational, sexy, driving, provocative, nostalgic, sentimental, . . ."—a challenging request, especially considering that the musical snippet was to be "not more than 3.8 seconds long."[4]

For brief function sounds, designers might create a click, beep, or buzz. They often start with a recording of something natural and then manipulate it in software. The audio processing might make the end result almost unrecognizable, but starting with the recording of a real sound gives the end result a natural aural complexity that affords it a sense of believability. The "unlock" sound of the iPhone is very similar to the click and spring of a locking pliers opening. Function sounds are best when they fit with the dimensions of the digital device, using frequencies that could plausibly be created from a mechanical object of a similar size. When sounds and functions are properly matched, the electronic object starts to feel mechanical.[5]

It makes me uneasy that our aural environment is becoming peppered with sounds primarily designed to sell products. Globalization of technologies also causes an attendant homogenization of the noises that form the backing track to our lives. Although electronic products allow sounds to be changed and customized, a sonic free-for-all is not always a good idea. I remember the cacophony created by personalized ringtones, which are fortunately now less fashionable. I would suggest that "personalization" should be done only en masse, using something that resonates with local culture and history. Maybe electronic cars in Bangkok could re-create the putt-putt of the tuk-tuk rickshaws, and the population of Manchester could change its ringtones to play the clatter of the cotton mills that transformed the city during the Industrial Revolution.

If I look back a couple of decades from now, some of today's technical sounds will be nostalgic sonic wonders. I can be sure of this because it has happened before. When I hear the two-tone beeps from *Pong*, I am reminded of playing that computer game at a friend's house as a teenager. As we've learned from people's reactions to birdsong, sonic nostalgia will not just be about the most unusual or aesthetically pleasing; it will include everyday sounds associated with strong individual memories. In the future, maybe couples will not just have "our tune" but "our bleep"; they'll cherish the alert sound as their loved one's message arrives on Facebook.

What first drew me to architectural acoustics is the fusion of the objectivity of physics with the subjectivity of perception. Engineers might have sophisticated computer programs to model the physics of sound waves, but that counts for nothing if listeners judge the acoustics to be poor and find what they hear unacceptable. In a grand concert hall, the audience judges whether their enjoyment is enhanced by the room's acoustic. In a noisy school canteen, the students are annoyed that they cannot easily chat with friends. Scientists have worked out the physiology of this hearing, but we only partially understand how a brain then processes and emotionally responds to sound. Despite this chasm in knowledge, computer models are invaluable, enabling engineers to calculate how many acoustic absorbers are needed to quiet the school canteen or what shape to make a concert hall to enhance the music. Meanwhile, scientists are striving to take the models further and predict what happens between the ears.

In all the architectural sonic wonders I visited, it was what happened after the initial impulse from a balloon burst, hand clap, or gunshot that distinguished the most remarkable places, especially if it confounded my knowledge of physics. Psychologists and neuroscientists are just starting to unravel how expectation plays a crucial role in our response to sound. A common example is music, in which composers toy with our emotions by subverting what listeners anticipate. Scientists have tested this idea by measuring changes in skin conductance when a note or chord in a piece of music is altered to something surprising. Unexpected notes cause listeners to sweat a little more—physiological evidence of an emotional response.[6] Subverted expectation was important to my perception of the gunshot in the oil storage tank at Inchindown (see Chapter 1). I had anticipated a long reverberation time, but I was astonished by the tsunami of sound that enveloped me and took a ludicrously long time to die away. Open a book on architectural acoustics and look at a table of reverberation times in classrooms, concert halls, and cathedrals, and none of them will have values close to what I measured at Inchindown. In this hidden complex, deep within a hillside, I felt like a gentleman explorer from a century ago. There was the claustrophobic entrance to the oily concrete cavern through the tight pipework, the revelation of the awe-inspiring sound, and of course, the feeling of uniqueness: no one had tested acoustics like this before.

I have discovered that dilapidated buildings, abandoned military installations, and the remnants of industry offer some of the most unusual acoustics. The disused cooling towers of the Thorpe Marsh Power Station in England was the one sonic wonder that eluded me. The station closed in 1994, but the tall, hourglass brick towers were left standing. According to the e-mail correspondent who suggested I should visit the decommissioned power plant, the 100-meter-high (330-foot) towers produced "terrific" echoes inside. Even better, said my correspondent, there was no security, so one could just walk into the site from a neighboring road. One autumn day after I had finished visiting all the other sonic wonders, I packed up my recording gear and drove over to the site. Having been in the radome of Teufelsberg, I could guess the sound effects I might hear in the cooling towers: focused echoes in the center reverberating above my head and whispering-gallery effects around the edge were likely. I took along my saxophone because I thought it would be fun to try and improvise with the echo, and see how that might affect me as a performer.

But, disaster! All that remained of the towers were some very large piles of rubble. After standing untouched for eighteen years, they had been torn down a month earlier. As I drove away disappointed, I was reminded of the story of the Teatro La Fenice in Venice, one of the best-sounding opera houses in the world. The building burned down in 1996. Fortunately, two months before the fire, binaural recordings of the theater had been made. Binaural measurements use a mannequin with microphones buried in the side of its head to pick up what would normally pass down the ear canals of a listener. Unlike normal stereo, listening back to these types of recordings can give you a real sense of being in a space. The binaural recordings of the Venice opera house helped inform the reconstruction.[7]

Efforts to document acoustics have focused on auditoriums, churches, and ancient sites like Stonehenge. Recording the acoustic footprint preserves it for posterity, and makes it possible to bring a place back to life by rendering it in virtual reality. But we should also be preserving the remarkable acoustics of more modern places. There are three radomes on the roof of the disused listening station at Teufelsberg in Berlin, but two of these have been very badly vandalized. Will someone capture the acoustic signature of the last radome before it, too, becomes damaged and the sound is lost forever? Heritage organizations need to realize the importance of sound and not just document sites in words and pictures. There must also be other detritus of human progress harboring sonic wonders waiting to be discovered. And no doubt, new constructions are unwittingly producing the sonic wonders of the future as I write.

While this book is about finding the most remarkable sounds, I have noticed that keeping an ear out for extraordinary examples has made me enjoy and take more notice of everyday sounds. It was in the Mojave Desert that I first really noticed how evergreen trees whistle. Now, as I walk near my home, I listen for the rustling plane trees lining the streets, and I can even enjoy the wind whistling through that scourge of suburbia, the Leyland cypress. One morning I got up very early to hear bitterns boom because they make the oddest bird call in Britain, and now I listen for snatches of birdsong as I cycle to work dodging between cars. I now appreciate how varied water sounds can be, from the overwhelming roar of the Dettifoss waterfall to the more subtle babbling brook in my local city park.

There must be other natural sonic wonders waiting to be heard by humans. Every week, new animal species are discovered, and since nearly every one of them hears sound or senses vibration, new animal calls are bound to be revealed. It is a great time for amateur naturalists to seek out and record such sounds. The recording of audio on video cameras or mobile phones is becoming increasingly common. Many of us now carry around technology that can capture sonic wonders and share them with friends and family. New natural behaviors, such as the way one vine has evolved to attract pollinating bats (described in Chapter 3)—new ways that animals and plants exploit sound—will also be discovered.

At my university, the anechoic chamber wows visitors because the silence allows them to listen to their hearts and minds. I have always thought it would be good to have one in a shopping mall so that more of the public could experience the silence. I think it would also be fun to make one with transparent walls; at least one concert hall has been built with huge glass walls, so why not an anechoic chamber? To do this, the foam wedges that cover every interior surface of a conventional design would have to be replaced with transparent absorbers. There has been a lot of interest in transparent acoustic treatments lately because they fit with the trend in architecture for lots of glazing. They can be made from perforated plastic, a bit like the rustling plastic bags that warm bread used to be sold in. They are not perfect at removing sound, but by curving the walls of a see-through anechoic chamber, like the bottom half of a goldfish bowl, any reflected sound could be directed up above the head of the listeners. In this room you could take a break from the city in complete silence, while others walk by laden with shopping bags.

For me, the more conventional anechoic chamber at Salford University has become a normal room for scientific experimentation, partly because of my brain's automatic habituation to the acoustic, but also because I took it for granted. I started collecting sonic wonders because I realized I needed to rediscover the skill of listening. In an effort to awaken my ears, I went on soundwalks, participated in a silent retreat, and floated about in brine. Along the way I have had the chance to interview inspiring artists, sound recordists, and musicians who have demonstrated enviable sensitivity and understanding of the aural. They taught me so much, and made me realize that scientists and engineers need to listen more to them and to the world around us. I hope all of us will open our ears to the strange sounds around us. As my search draws to an end, I realize that I have changed. If we all listened to and cared for the sonic wonders around us, as I now try to do, we would start to build a better-sounding world.

Acknowledgments

It has been a privilege to discuss acoustics with many great people over the last twenty-five years. I would like to thank the following, who, for this book, explained acoustic phenomena to me or helped me to gain new sound experiences: Keith Attenborough; Mark Avis; Michael Babcock; Barry Blesser; David Bowen; Stuart Bradley; Andrew Brookes; Angus Carlyle; Mike Caviezel; Dominic Chennell; Rob Connetta; Frances Crow; Marc Crunelle; John Culling; Peter Cusack; Helen Czerski; Peter D'Antonio; Bill Davies; Charles Deenen; Stéphane Douady; John Drever; Bruno Fazenda; Linda Gedemer; Tim Gedemer; Tony Gibbs; Wendy Hasenkamp; Marc Holderied; Diane Hope; Seth Horowitz; Simon Jackson; Brian Katz; Paul Kendrick; Allan Kilpatrick; Tim Leighton; Jane MacGregor; Katherine MacLean; Paul Malpas; Barry Marshall; Henric Mattsson; Bryony McIntyre; Daniel Mennill; Andy Moorhouse; Myron Nettinga; Stuart Nolan; James Pask; Lee Patterson; Chris Plack; Eleanor Ratcliffe; Brian Rife; John Roesch; Duncan from the Royal Society for the Protection of Birds (RSPB); Martin Schaffert; Ann Scibelli; Clare Sefton; Jonathan Sheaffer;

Bridget Shield; Matt Stephenson; Davide Tidoni; Rupert Til; Lamberto Tronchin; Rami Tzabar; Nathalie Vriend; Chris Watson; Nick Whitaker; Andrew Whitehouse; Heather Whitney; Pascal Wyse; Luray Caverns; members of Subterranea Britannica; the teachers, coordinators, fellow retreatants, and staff at the Buddhist retreat; and anyone else I accidentally left off this list.

I thank the Engineering and Physical Sciences Research Council for my Senior Media Fellowship, which gave me the time to develop the proposal for this book. Also many have helped me to develop as a science communicator, including staff at the BBC Radio Science Unit and *New Scientist*.

My agent, editors, and copy editor have been highly influential in shaping the overall narrative of this book and improving the writing. I'm indebted to Stephanie Hiebert, Tom Mayer, Zoë Pagnamenta, Kay Peddle, Peter Tallack, and Gemma Wain.

Thank you to Nathan Cox, who helped with some of the diagrams. Finally, thanks to the following, who commented on early drafts: Deborah, Jenny, Peter, and Stephen Cox.

Notes

Prologue

1   M. Spring, "Bexley Academy: Qualified Success," *Building*, June 12, 2008.

2   *The New Yorker* described *The Phantom Tollbooth* as the "closest thing that American literature has to an 'Alice in Wonderland.'" A. Gopnik, "Broken Kingdom: Fifty Years of the 'Phantom Tollbooth,'" *New Yorker*, October 17, 2011, http://www.newyorker.com/reporting/2011/10/17/111017fa_fact_gopnik.

3   British Library Sounds, "Programme II: B—Part 1: Listening. Soundscapes of Canada," http://sounds.bl.uk/View.aspx?item=027M-W1CDR0001255 -0200V0.xml, accessed October 6, 2011.

4   R. M. Schafer, *The Soundscape: Our Sonic Environment and the Tuning of the World* (Rochester, VT: Destiny Books, 1994), 208.

5   Bill Davies, personal communication, September 2011.

6   C. Spence and V. Santangelo, "Auditory Attention," in *The Oxford Handbook of Auditory Science: Hearing*, ed. C. J. Plack (Oxford: Oxford University Press, 2010). When not writing about auditory attention, Charles Spence researches how sound affects taste.

7   Switching talkers works only if both speakers are the same gender.

8   UK data from MORI Social Research Institute: *Neighbour Noise: Public Opinion Research to Assess Its Nature, Extent and Significance* (Department for Environment, Food and Rural Affairs, 2003). US data from the 2000 US

census as reported in L. Goines and L. Hagler, "Noise Pollution: A Modern Plague," *Southern Medical Journal* 100 (2007): 287–94. EU data from *Future Noise Policy*, European Commission Green Paper, COM (96) 540 final (Brussels: Commission of the European Communities, 1996).

9   V. J. Rideout, U. G. Foehr, and D. F. Roberts, *Generation M²: Media in the Lives of 8- to 18-Year-Olds* (Menlo Park, CA: Kaiser Family Foundation, 2010).

10   The quote is from Mike Caviezel (personal communication, May 13, 2011), whose experiences are given in full later in the book.

11   R. Campbell-Johnston, "Hockney Works Speak of Rapture," *Times* (London), January 21, 2012.

1: The Most Reverberant Place in the World

1   A. Tajadura-Jiménez, P. Larsson, A. Väljamäe, D. Västfjäll, and M. Kleiner, "When Room Size Matters: Acoustic Influences on Emotional Responses to Sounds," *Emotion* 10 (2010): 416–22.

2   *Encyclopaedia Britannica*, "Wallace Clement Sabine," http://www.britannica.com/EBchecked/topic/515073/Wallace-Clement-Sabine, accessed May 30, 2013.

3   W. C. Sabine, "Architectural Acoustics: Correction of Acoustical Difficulties," *Architectural Quarterly of Harvard University*, March 1912.

4   R. T. Beyer, *Sounds of Our Times: Two Hundred Years of Acoustics* (New York: Springer, 1998). Original appeared in H. Matthews, *Observations on Sound* (publisher unknown, 1826).

5   The physical volume might be altered as well. In designing a concert hall for classical music, at least 10 cubic meters (350 cubic feet) per seat is a useful rule of thumb.

6   This quote comes from 1972, just before the lecture hall was improved by further renovations. B. F. G. Katz and E. A. Wetherill, "Fogg Art Museum . . . Room Acoustics" (paper presented at Forum Acusticum, Budapest, Hungary, August 29–September 2, 2005). The hall was demolished in 1973 to make way for student accommodation.

7   The value is for a full audience at midfrequency.

8   The quote comes from L. L. Beranek, *Music, Acoustics & Architecture* (Hunting, NY: Krieger, 1979), which includes a wonderful first chapter detailing some of the myths of concert hall acoustics.

9   P. Doyle, *Echo and Reverb: Fabricating Space in Popular Music*, 1900–1960 (Middletown, CT: Wesleyan University Press), 143.

10 M. Barron, *Auditorium Acoustics and Architectural Design*, 2nd ed. (London: Spon Press/Taylor & Francis, 2010), 103.

11 G. A. Soulodre, "Can Reproduced Sound Be Evaluated Using Measures Designed for Concert Halls?" (paper presented at Spatial Audio & Sensory Evaluation Techniques Workshop, Guildford, UK, April 6–7, 2006).

12 *All My Children* was a soap opera that appeared on ABC for forty-one years. The quote is from J. C. Jaffe, *The Acoustics of Performance Halls* (New York: W. W. Norton, 2010).

13 Other changes also affected the acoustics. See L. L. Beranek, "Seeking Concert Hall Acoustics," *IEEE Signal Processing Magazine*, 24 (2007): 126–30.

14 Ibid.

15 Loudness is also important for this effect. As an orchestra plays louder, for example, the envelopment and broadening increase.

16 Brian Eno, speaking on the BBC Radio 4 program *Acoustic Shadows*, broadcast September 14, 2004.

17 This quote comes from *Concert Hall Acoustics: Art and Science*, a 2001 exhibit at the South Bank Centre, London. Source unknown.

18 L. L. Beranek, *Concert Halls and Opera Houses*, 2nd ed. (New York: Springer, 2004), 7–8.

19 Barron, *Auditorium Acoustics*, 153.

20 S. Quinn, "Rattle Plea for Bankrupt Orchestras," *Guardian* (London), July 13, 1999.

21 D. Trevor-Jones, "Hope Bagenal and the Royal Festival Hall," *Acoustics Bulletin* 26 (May 2011): 18–21.

22 It was an underestimation of the audience absorption that is mostly to blame for the lack of reverberance in the hall; see B. M. Shield, "The Acoustics of the Royal Festival Hall," *Acoustics Bulletin* 26 (May 2011): 12–17.

23 R. A. Laws and R. M. Laws, "Assisted Resonance and Peter Parkin," *Acoustics Bulletin* 26 (May 2011): 22–29.

24 I found reverberation time values for the Taj Mahal on the web, with figures ranging from 10 to 30 seconds, but I could not find a reliable source. Similarly, there are references to a reverberation time of 20 seconds for Gol Gumbaz, but no information to check the provenance of the number.

25 Tor Halmrast, personal communication, October 3, 2011.

26 This reverberation time comes from A. Buen, "How Dry Do the Record-

ings for Auralization Need to Be?" *Proceedings of the Institute of Acoustics* 30 (2008): 108; it was recorded when there were twenty-five people in the room. A better value for the empty room at midfrequency is about 11 seconds, estimated using an impulse response from the software Altiverb.

27   Or one needs to use a measure such as reverberation time, which does not depend on the loudness of the initial sound.

28   The poem is from Hilaire Belloc's *Cautionary Tales for Children*, first published in 1907, and begins, "A trick that everyone abhors, / In little girls is slamming doors." Rebecca gets her comeuppance when a bust falls from above a door and kills her.

29   For the audio geeks out there, this is an average of the 500-, 1,000-, and 2,000-hertz octave bands. The calculation is based on measurement data from Damian Murphy from the University of York (http://www.openairlib .net/auralizationdb/content/hamilton-mausoleum, accessed July 15, 2012) because there were too many people in the room (who absorb sound) when I was visiting.

30   See P. Darlington, "Modern Loudspeaker Technology Meets the Medieval Church," *Proceedings of the Institute of Acoustics*, 2002; or the modestly titled paper by D. Lubman and B. H. Kiser, "The History of Western Civilization Told through the Acoustics of Its Worship Spaces" (paper presented at the 19th International Congress on Acoustics, Madrid, September 2–7, 2007).

31   R. C. Rath, "Acoustics and Social Order in Early America," in *Hearing History: A Reader*, ed. M. M. Smith (Athens: University of Georgia Press, 2004), 209.

32   This estimate is based on Barron, *Auditorium Acoustics*, 19.

33   S. J. van Wijngaarden and R. Drullman, "Binaural Intelligibility Prediction Based on the Speech Transmission Index," *Journal of the Acoustical Society of America* 123 (2008): 4514–23. The phenomenon is similar to what happens with the cocktail party effect, that magical process that enables us to pick out the sound of a single speaker from the hubbub of the rest of a party.

34   Even if the priest is not straight ahead, these are ways for the brain to exploit this binaural processing.

35   For a nonecclesiastical example, see H. M. Goddard, "Achieving Speech Intelligibility at Paddington Station," *Journal of the Acoustical Society of America* 112 (2002): 2418. The acoustic principles are the same.

36   P. F. Smith, *The Dynamics of Delight: Architecture and Aesthetics* (London: Routledge, 2003), 21.

37   Beranek, Concert Halls and Opera Houses, 9. Incidentally, Richard Wagner

was an example of a composer becoming a successful acoustician when he helped design the Bayreuth Festival Theatre in 1876. The innovative orchestral pit had space for up to 130 players and extended deep under the stage. Since there is no direct line of sight between orchestra and audience, much of the treble sound is lost. In addition to creating the distinctive subdued, haunting Wagner sound, this arrangement enables the singers to be heard above a large orchestra.

38 T. H. Lewers and J. S. Anderson, "Some Acoustical Properties of St. Paul's Cathedral, London," *Journal of Sound and Vibration* 92 (1984): 285–97.

39 F. Jabr, "Gunshot Echoes Used to Map Caves' Interior," *New Scientist*, no. 2815 (June 9, 2011): 26.

40 R. Newmarch, *The Concert Goer's Library of Descriptive Notes* (Manchester, NH: Ayer, 1991), 72.

41 This measurement is featured in "Western Isles and Shetland," which was the fourth episode of the sixth series of the BBC television program *Coast*, first broadcast July 3, 2011.

42 B. Blesser and L.-R. Salter, *Spaces Speak, Are You Listening?: Experiencing Aural Architecture* (Cambridge, MA: MIT Press, 2007), 180.

43 Quotes from *Resonant Spaces*, "What's It All About?" http://arika.org.uk/resonant-spaces/what/?, accessed July 23, 2012; and from a poster advertising the tour in the offices of James Pask.

44 Mike Caviezel, personal communication, May 13, 2011.

45 These dimensions are very rough estimates from my visit.

46 This contrasts starkly with the American reservoir and how it was described to me by Mike Caviezel. Wormit is a cuboid, whereas the American reservoir is a huge cylinder, which might cause focusing and explain the different perceived qualities.

47 The exact reason for this difference is unclear, but the impulse responses I measured with a balloon suggest that there are a lot of early beneficial reflections in the reservoir.

48 W. Montgomery, "WIRE Review of Resonant Spaces," *Wire* 299 (January 2009).

49 Ibid.

50 Over the trombone frequency range, the time taken to drop 10 decibels is 3 seconds, based on a more accurate estimation of reverberation time given later in the chapter.

51 "Album Reviews," *Billboard*, September 16, 1995.

52 D. Craine, "Strangeness in the Night," *Times* (London), November 16, 2001.

53 "Stuart Dempster Speaks about His Life in Music: Reflections on His Fifty Year Career as a Trombonist, in Conversation with Abbie Conant," http://www.osborne-conant.org/Stu_Dempster.htm, accessed July 19, 2012.

54 This is for a strange frequency range of 125–2,500 hertz, dictated by the instruments on the recording. The extraction method is described in P. Kendrick, T. J. Cox, F. F. Li, Y. Zhang, and J. A. Chambers, "Monaural Room Acoustic Parameters from Music and Speech," *Journal of the Acoustical Society of America* 124 (2008): 278–87. The 27-second value is an overestimation because in such a reverberant space, with multiple players layering sound, it is hard to tell when the musicians stop playing notes.

55 W. D. Howells, *Italian Journeys* (publisher unknown, 1867), 233.

2: Ringing Rocks

1 A number of papers exploring the acoustics of ancient sites have been met with skepticism. The one I have in mind is R. G. Jahn, P. Devereux, and M. Ibison, "Acoustical Resonances of Assorted Ancient Structures," *Journal of the Acoustical Society of America* 99 (1996): 649–58.

2 J. L. Stephens, *Incidents of Travel in Greece, Turkey, Russia and Poland* (Edinburgh: William and Robert Chambers, 1839), 21.

3 M. Barron, *Auditorium Acoustics and Architectural Design*, 2nd ed. (London: Spon Press/Taylor & Francis, 2010), 276.

4 B. Thayer, "Marcus Vitruvius Pollio: de Architectura, Book V" [translation], http://penelope.uchicago.edu/Thayer/E/Roman/Texts/Vitruvius/5*.html, accessed October 18, 2011.

5 This is why schoolteachers keep reminding their pupils to turn and talk to the audience when performing for their parents at assemblies.

6 Barron, *Auditorium Acoustics*, 277.

7 E. Rocconi, "Theatres and Theatre Design in the Graeco-Roman World: Theoretical and Empirical Approaches," in *Archaeoacoustics*, ed. C. Scarre and G. Lawson (Cambridge: McDonald Institute for Archaeological Research, 2006), 72.

8 A number of academics have tried to decode the development of theaters to look for acoustic understanding, including J. Kang and K. Chourmouziadou in "Acoustic Evolution of Ancient Greek and Roman Theatres," *Applied Acoustics* 69 (2008): 514–29.

9 B. Blesser and L.-R. Salter, *Spaces Speak, Are You Listening?: Experiencing Aural Architecture* (Cambridge, MA: MIT Press, 2007). A great book setting out how architectural acoustics affects us.

10 Vitruvius also advocated the use of resonating vases to detect assailants digging tunnels under the walls of Apollonia. The bronze vessels were hung from the ceiling, and the blows of the excavators' tools excited the resonance, according to F. V. Hunt, *Origins in Acoustics* (New Haven, CT: Yale University Press, 1978), 36.

11 Thayer, "Marcus Vitruvius Pollio."

12 M. Kayili, *Acoustic Solutions in Classic Ottoman Architecture* (Manchester, UK: FSTC Limited, 2005).

13 A. P. O. Carvalho, V. Desarnaulds, and Y. Loerincik, "Acoustic Behavior of Ceramic Pots Used in Middle Age Worship Spaces—A Laboratory Analysis" (paper presented at the 9th International Congress on Sound and Vibration, Orlando, FL, July 8–11, 2002).

14 P. V. Bruel, "Models of Ancient Sound Vases," *Journal of the Acoustical Society of America* 112 (2002): 2333. Quite a few academics have made measurements on vases and reached similar conclusions.

15 L. L. Beranek, *Music, Acoustics and Architecture* (New York: Wiley, 1962), 5.

16 D. Richter, J. Waiblinger, W. J. Rink, and G. A. Wagner, "Thermoluminescence, Electron Spin Resonance and ^{14}C-Dating of the Late Middle and Early Upper Palaeolithic Site of Geissenklösterle Cave in Southern Germany," *Journal of Archaeological Science* 27 (2000): 71–89. This work made headlines around the world, so information can also be found on news websites—for example, P. Ghosh, "'Oldest Musical Instrument' Found," *BBC News*, June 25, 2009, http://news.bbc.co.uk/1/hi/8117915.stm.

17 F. d'Errico and G. Lawson, "The Sound Paradox," in *Archaeoacoustics*, ed. C. Scarre and G. Lawson (Cambridge: McDonald Institute for Archaeological Research, 2006), 50.

18 I. Morley, *The Evolutionary Origins and Archaeology of Music*, Darwin College Research Report DCRR-002 (Cambridge: Darwin College, Cambridge University, 2006).

19 N. Boivin, "Rock Art and Rock Music: Petroglyphs of the South Indian Neolithic," *Antiquity* 78 (2004): 38–53.

20 L. Dams, "Palaeolithic Lithophones: Descriptions and Comparisons," *Oxford Journal of Archaeology* 4 (1985): 31–46.

21 Luray Caverns, "Discovery," http://luraycaverns.com/History/Discovery/tabid/529/Default.aspx, accessed June 17, 2012.

22 H. H. Windsor, "The Organ That Plays Stalactite," *Popular Mechanics*, September 1957.

23 An early press report suggested, "When four-year old Robert Sprinkle bumped his head on a stalactite while visiting the Caverns in June 1954, the deep reso-

nant tone of the rock fascinated him and his father." "Stalactite Organ Makes Debut," *Pittsburgh Post-Gazette*, June 9, 1957. A great story, but according to the press office at Luray Caverns, it is sadly incorrect.

24 "The Rock Hamonicon," *Journal of Civilization* (1841).

25 Ibid.

26 Ibid.

27 J. Blades, *Percussion Instruments and Their History* (London: Kahn & Averill, 2005), 90.

28 Allerdale Borough Council, "The Musical Stones of Skiddaw: The Richardson Family and the Famous Musical Stones of Skiddaw," http://www.allerdale .gov.uk/leisure-and-culture/museums-and-galleries/keswick-museum/the -musical-stones-of-skiddaw.aspx, accessed March 15, 2011.

29 Blades, *Percussion Instruments*, 90.

30 M. Wainwright, "Evelyn Glennie's Stone Xylophone," *Guardian* (London), August 19, 2010.

31 "Online Special: Ruskin Rocks!" *Geoscientist Online*, October 4, 2010, http:// www.geolsoc.org.uk/ruskinrocks, accessed May 16, 2011.

32 Hunt, *Origins in Acoustics*, 152.

33 Strictly speaking, this description fits only when you are in the middle of the stairwell, equidistant from the walls.

34 The quote comes from the Tatton Park Biennial 2012 exhibition catalogue: http://www.tattonparkbiennial.org/detail/3070, accessed March 17, 2011.

35 I. Reznikoff, "On Primitive Elements of Musical Meaning," *Journal of Music and Meaning* 3 (Fall 2004/Winter 2005).

36 S. J. Waller, "Sound and Rock Art," *Nature* 363 (1993): 501.

37 David Lubman, personal communication, June 25, 2012.

38 S. J. Waller, speaking on the BBC Radio 4 program *Acoustic Shadows*, broadcast September 14, 2004. I also appeared on this program, talking about acoustic test chambers, such as the anechoic chamber described in Chapter 7.

39 L. Dayton, "Rock Art Evokes Beastly Echoes of the Past," *New Scientist*, no. 1849 (November 28, 1992), 14.

40 Waller, *Acoustic Shadows* radio broadcast.

41 Dayton, "Rock Art."

42 This was the second trip that failed acoustically. When I tried to visit the Rouffignac Cave in France, I found that the floor had been excavated to fit an electric train into the cavern, which altered the acoustics and rendered any sonic investigation pointless.

43   D. Wilson, *Hiking Ruins Seldom Seen: A Guide to 36 Sites across the Southwest*, 2nd ed. (Guilford, CT: Falcon Guides, 2011), 16–17. The book reports on a study by archaeologist Donald E. Weaver, who dates the petroglyphs to somewhere between AD 900 and 1100.

44   Ibid.

45   P. Schaafsma, "Excerpts from *Indian Rock Art of the Southwest*," in *The Archeology of Horseshoe Canyon* (Moab, UT: Canyonland National Park, date unknown), 15.

46   Waller, "Sound and Rock Art."

47   Lubman, personal communication, June 25, 2012.

48   The pyramid is 24 meters (80 feet) high, and the square base is 56 meters (185 feet) wide.

49   A colleague of mine has suggested that the corrugated roof might also chirp.

50   Lubman, personal communication, June 25, 2012.

51   C. Scarre, "Sound, Place and Space: Towards an Archaeology of Acoustics," in *Archaeoacoustics*, ed. C. Scarre and G. Lawson (Cambridge: McDonald Institute for Archaeological Research, 2006), 6.

52   T. Hardy, *Tess of the D'Urbervilles* (Rockville, MD: Serenity Publishers, 2008), 326.

53   D. Barrett, "Review: Collected Works," *New Scientist*, no. 2118 (January 24, 1998), 45.

54   It is not known whether Stonehenge was used for sacrifices. The real purpose of the site is still open to debate.

55   Bruno Fazenda, personal communication, October 2011.

56   Ibid.

57   P. Devereux, *Stone Age Soundtracks: The Acoustic Archaeology of Ancient Sites* (London: Vega, 2001), 103.

58   It is possible to get echoes from rings of columns. The Court of 3 Stars at Bicentennial Capitol Mall State Park in Nashville, Tennessee, has fifty 7.6-meter-high (25-foot) limestone columns arranged in two C shapes facing each other. The columns contain bells to play the "Tennessee Waltz" every fifteen minutes. I have been told these columns produce a distinctive echo if you stand in the middle of the circle, and videos on the Internet seem to confirm this claim.

59   Jahn, Devereux, and Ibison, "Acoustical Resonances."

60   Devereux, *Stone Age Soundtracks*, 86–89.

61   I became aware of this paper from discussions with Matthew. A lay-language version of his paper is available at http://www.acoustics.org/press/153rd/wright.html, accessed October 25, 2011.

62   Matthew Wright could also have asked whether a burial mound is like a car. Some years ago I was solicited to produce pseudoscience for an advertising campaign proclaiming which model of car was best for singing in. I declined the offer.

3: Barking Fish

1   Loss of food and habitats caused by intensive farming is probably more likely to blame for the loss of songbirds. A study by the British Trust for Ornithology (BTO) found that songbird numbers were no different in places where there were many magpies and places where there were few. S. E. Newson, E. A. Rexstad, S. R. Baillie, S. T. Buckland, and N. J. Aebischer, "Population Changes of Avian Predators and Grey Squirrels in England: Is There Evidence for an Impact on Avian Prey Populations?" *Journal of Applied Ecology* 47 (2010): 244–52.

2   Chris Watson, personal communication, September 15, 2011.

3   R. S. Ulrich, "View through a Window May Influence Recovery from Surgery," *Science* 224 (1984): 420–21. Other studies have shown that nature reduces stress for office workers and prisoners; see G. N. Bratman, J. P. Hamilton, and G. C. Daily, "The Impacts of Nature Experience on Human Cognitive Function and Mental Health," *Annals of the New York Academy of Sciences* 1249 (2012): 118–36.

4   The beneficial effect of nature even happens when people are just shown photos rather than experiencing nature firsthand; see "A Walk in the Park a Day Keeps Mental Fatigue Away," *Science News*, December 23, 2008, http://www.sciencedaily.com/releases/2008/12/081218122242.htm.

5   R. S. Ulrich, R. F. Simons, B. D. Losito, E. Fiorito, M. A. Miles, and M. Zelson, "Stress Recovery during Exposure to Natural and Urban Environments," *Journal of Environmental Psychology* 11 (1991): 201–30.

6   J. J. Alvarsson, S. Wiens, and M. E. Nilsson, "Stress Recovery during Exposure to Nature Sound and Environmental Noise," *International Journal of Environmental Research and Public Health* 7 (2010): 1036–46. The study found no effect for heart rate.

7   S. Kaplan and R. Kaplan, *The Experience of Nature: A Psychological Perspective* (New York: Cambridge University Press, 1989).

8   Myron Nettinga, personal communication, June 24, 2012.

9   H. C. Gerhardt and F. Huber, *Acoustic Communication in Insects and Anu-*

rans: Common Problems and Diverse Solutions (Chicago: University of Chicago Press, 2002).

10   L. Elliot, *A Guide to Wildlife Sounds* (Mechanicsburg, PA: Stackpole Books, 2005), 86. For degrees Celsius, divide the number of chirps per minute by 7 and add 4.

11   A drop of an octave is a halving of the frequency. The first note is at about 1,300 hertz, somewhere in the middle of the piccolo's range.

12   P. C. Nahirney, J. G. Forbes, H. D. Morris, S. C. Chock, and K. Wang, "What the Buzz Was All About: Superfast Song Muscles Rattle the Tymbals of Male Periodical Cicadas," *FASEB Journal* 20 (2006): 2017–26.

13   Nettinga, personal communication, June 24, 2012.

14   M. Stroh, "Cicada Song Is Illegally Loud," *Baltimore Sun*, May 16, 2004.

15   Ibid.

16   While *Micronecta scholtzi* is the loudest known aquatic creature relative to its size, comparison with land animals is best avoided. Although the media have claimed that the insect "reached 78.9 decibels, comparable to a passing freight train," this comparison is incorrect; see T. G. Leighton, "How Can Humans, in Air, Hear Sound Generated Underwater (and Can Goldfish Hear Their Owners Talking)?" *Journal of the Acoustical Society of America* 131 (2012): 2539–42. The original research published—J. Sueur, D. Mackie, and J. F. C. Windmill, "So Small, So Loud: Extremely High Sound Pressure Level from a Pygmy Aquatic Insect (Corixidae, Micronectinae)," *PLoS One* 6 (2011): e21089—did not allow for the difference between the density and speed of sound in air versus water.

17   J. Theiss, "Generation and Radiation of Sound by Stridulating Water Insects as Exemplified by the Corixids," *Behavioral Ecology and Sociobiology* 10 (1982): 225–35.

18   M. Versluis, B. Schmitz, A. von der Heydt, and D. Lohse, "How Snapping Shrimp Snap: Through Cavitating Bubbles," *Science* 289 (2000): 2114–17.

19   Watson, personal communication, September 15, 2011.

20   Ibid.

21   See "Snapping Shrimp Drown Out Sonar with Bubble-Popping Trick, Described in *Science*," *Science News*, September 22, 2000, http://www.sciencedaily.com/releases/2000/09/000922072104.htm.

22   D. Livingstone, *Missionary Travels and Researches in South Africa* (London: J. Murray, 1857).

23   The actual value measured was 106.7 decibels at a distance of 50 centimeters

(about 20 inches), but I have estimated a 1-meter value to be consistent with the rest of the chapter. J. M. Petti, "Loudest," in *University of Florida Book of Insect Records*, ed. T. Walker, chap. 24 (Gainesville: University of Florida, 1997), http://entnemdept.ufl.edu/walker/ufbir/chapters/chapter_24.shtml.

24   Livingstone, *Missionary Travels*.

25   Which bit vibrates and radiates can depend on the species. In bullfrogs, tympanic membranes are important; see A. P. Purgue, "Tympanic Sound Radiation in the Bullfrog *Rana catesbeiana*," *Journal of Comparative Physiology. A, Sensory, Neural, and Behavioral Physiology* 181 (1997): 438–45.

26   A. S. Rand and R. Dudley, "Frogs in Helium: The Anuran Vocal Sac Is Not a Cavity Resonator," *Physiological Zoology* 66 (1993): 793–806.

27   M. J. Ryan, M. D. Tuttle, and L. K. Taft, "The Costs and Benefits of Frog Chorusing Behavior," *Behavioral Ecology and Sociobiology* 8 (1981): 273–78.

28   J. Treasure, "Shh! Sound Health in 8 Steps," *TED Talks*, September 2010, http://www.ted.com/talks/julian_treasure_shh_sound_health_in_8_steps .html, accessed September 20, 2012. Musicologist Joseph Jordania suggests that we find humming relaxing because silence is a sign of danger; J. Jordania, "Music and Emotions: Humming in Human Prehistory," in *Problems of Traditional Polyphony. Materials of the Fourth International Symposium on Traditional Polyphony, held at the International Research Centre of Traditional Polyphony at Tbilisi State Conservatory on September* 15-19, 2008, ed. R. Tsurtsumia and J. Jordania (Tbilisi, Republic of Georgia: Nova Science, 2010), 41–49.

29   J. Letzing, "A California City Is into Tweeting—Chirping, Actually—in a Big Way," *Wall Street Journal*, January 17, 2012.

30   B. Manilow, "Barry's Response to Australia's Plan," the *BarryNet*, July 18, 2006, http://www.barrynethomepage.com/bmnet000_060718.shtml.

31   The quotes in this section are all taken from Andrew Whitehouse's research blog, *Listening to Birds*, http://www.abdn.ac.uk/birdsong/blog, accessed September 2, 2012.

32   Actually, if you compare recordings of a whip and a whipbird side by side, they are quite different. The real whip creates a crack via a sonic boom and does not have a glissando, or the starting tone.

33   The glissando can also go in the opposite direction, from high to low frequency.

34   Daniel Mennill discusses fitness in a paper with Amy Rogers—D. J. Mennill and A. C. Rogers, "Whip It Good! Geographic Consistency in Male

Songs and Variability in Female Songs of the Duetting Eastern Whipbird *Psophodes olivaceus*," *Journal of Avian Biology* 37 (2008): 93–100—but in an e-mail to me, Daniel pointed out that the fitness hypothesis is conjecture. By playing recordings of the calls to males and females and watching how the birds responded, Amy Rogers confirmed the duet hypothesis: A. C. Rogers, N. E. Langmore, and R. A. Mulder, "Function of Pair Duets in the Eastern Whipbird: Cooperative Defense or Sexual Conflict?" *Behavioral Ecology* 18 (2007): 182–88. Duets are probably used for defending territories as well.

35 The decibel levels for the birds come from M. Wahlberg, J. Tougaard, and B. Møhl, "Localising Bitterns *Botaurus stellaris* with an Array of Non-linked Microphones," *Bioacoustics* 13 (2003): 233–45; those for the trumpet, from O. Olsson and D. S. Wahrolén, "Sound Power of Trumpet from Perceived Sound Qualities for Trumpet Players in Practice Rooms" (master's thesis, Chalmers University of Technology, Sweden, 2010).

36 A. C. Doyle, *The Hound of the Baskervilles* (Hertfordshire: Wordsworth Editions, 1999), 70.

37 TEDx events are local TED events, conferences designed to spread ideas and change the world. See http://www.ted.com.

38 Most detectors use beats (described in Chapter 8) to adjust sounds so that they're audible to humans.

39 Allowing for a shift in frequency, this analogy comes from A. van Ryckegham, "How Do Bats Echolocate and How Are They Adapted to This Activity?" December 21, 1998, http://www.scientificamerican.com/article.cfm?id=how-do-bats-echolocate-an.

40 While it can provide some protection, the reflex is not quick enough for sudden sounds such as explosions, and it fatigues if the loud sound lasts too long; see S. Gelfand, *Essentials of Audiology*, 3rd edition (New York: Thieme, 2009), 44.

41 Chris Watson, speaking on the BBC Radio 4 program *The Listeners*, broadcast February 28, 2013.

42 R. Simon, M. W. Holderied, C. U. Koch, and O. von Helversen, "Floral Acoustics: Conspicuous Echoes of a Dish-Shaped Leaf Attract Bat Pollinators," *Science* 333 (2011): 631–33.

43 Watson, personal communication, September 15, 2011.

44 Ibid.

45 These were probably social calls between dolphins rather than echolocation signals, which are usually at too high a frequency for humans to hear. See, for

example, "Bottlenose Dolphins: Communication & Echolocation," *SeaWorld/ Busch Gardens Animals*, http://www.seaworld.org/animal-info/info-books/ bottlenose/communication.htm, accessed September 14, 2012.

46   E. R. Skeate, M. R. Perrow, and J. J. Gilroy, "Likely Effects of Construction of Scroby Sands Offshore Wind Farm on a Mixed Population of Harbour *Phoca vitulina* and Grey *Halichoerus grypus* Seals," *Marine Pollution Bulletin* 64 (2012): 872–81.

47   E. C. M. Parsons, "Navy Sonar and Cetaceans: Just How Much Does the Gun Need to Smoke Before We Act?" *Marine Pollution Bulletin* 56 (2008): 1248–57.

48   I first saw this quote in D. C. Finfer, T. G. Leighton, and P. R. White, "Issues Relating to the Use of a 61.5 dB Conversion Factor When Comparing Airborne and Underwater Anthropogenic Noise Levels," *Applied Acoustics* 69 (2008): 464–71.

49   Even these conversions are controversial, and some argue that airborne and underwater decibel values should never be compared.

50   I saw the *New York Times* quote first in T. G. Leighton, "How Can Humans, in Air, Hear Sound Generated Underwater?". A jet engine typically makes 200 dB re 2×10^{-5} Pa at 1 meter. If the sonar is taken to make 233 dB re 1 $\times 10^{-6}$ Pa at 1 meter underwater, then this might crudely be converted to an airborne equivalent of $233 - 61.5 = 171.5$ dB re 2×10^{-5} Pa—actually quieter than one jet engine! See D. M. F. Chapman and D. D. Ellis, "The Elusive Decibel: Thoughts on Sonars and Marine Mammals," *Canadian Acoustics* 26 (1998): 29–31.

51   G. V. Frisk, "Noiseonomics: The Relationship between Ambient Noise Levels in the Sea and Global Economic Trends," *Scientific Reports* 2 (2012): 437.

52   R. M. Rolland, S. E. Parks, K. E. Hunt, M. Castellote, P. J. Corkeron, D. P. Nowacek, S. K. Wasser, and S. D. Kraus, "Evidence That Ship Noise Increases Stress in Right Whales," *Proceedings of the Royal Society of London. B, Biological Sciences* 279 (2012): 2363–68.

53   L. G. Ryan, *Insect Musicians & Cricket Champions: A Cultural History of Singing Insects in China and Japan* (San Francisco: China Books & Periodicals, 1996), XIII.

54   Watson, personal communication, September 15, 2011.

55   Phil Spector is famous for producing dense pop music in the 1960s from layers of musical instruments often playing in unison. A good example is the hit "Da Doo Ron Ron" by the Crystals.

56   Watson, personal communication, September 15, 2011.

57 E. Nemeth, T. Dabelsteen, S. B. Pedersen, and H. Winkler, "Rainforests as Concert Halls for Birds: Are Reverberations Improving Sound Transmission of Long Song Elements?" *Journal of the Acoustical Society of America* 119 (2006): 620–26.

58 H. Sakai, S.-I. Sato, and Y. Ando, "Orthogonal Acoustical Factors of Sound Fields in a Forest Compared with Those in a Concert Hall," *Journal of the Acoustical Society of America* 104 (1998): 1491–97. In another paper the authors looked at a bamboo forest and found a 1.5-second reverberation time.

59 H. Slabbekoorn, "Singing in the Wild: The Ecology of Birdsong," in *Nature's Music: The Science of Birdsong*, ed. P. Marler and H. Slabbekoorn (Amsterdam: Elsevier, 2004), 198.

60 E. P. Derryberry, "Ecology Shapes Birdsong Evolution: Variation in Morphology and Habitat Explains Variation in White-Crowned Sparrow Song," *American Naturalist* 174 (2009): 24–33.

61 H. Slabbekoorn and A. den Boer-Visser, "Cities Change the Song of Birds," *Current Biology* 16 (2006): 2326–31. Slabbekoorn and colleagues have also shown that great tits change their songs between quiet and noisy parts of the city, see *A Problem with Noise*, BBC Radio 4, broadcast August 20, 2009. Also see H. Brumm, "The Impact of Environmental Noise on Song Amplitude in a Territorial Bird," *Journal of Animal Ecology* 73 (2004): 434–40; and R. A. Fuller, P. H. Warren, and K. J. Gaston, "Daytime Noise Predicts Nocturnal Singing in Urban Robins," *Biology Letters* 3 (2007): 368–70.

62 Hans Slabbekoorn, speaking on the BBC Radio 4 program *A Problem with Noise*, broadcast August 20, 2009.

63 D. Stover, "Not So Silent Spring," *Conservation Magazine* 10 (January–March 2009).

64 D. Kroodsma, "The Diversity and Plasticity of Birdsong," in *Nature's Music: The Science of Birdsong*, ed. P. Marler and H. Slabbekoorn (Amsterdam: Elsevier, 2004), 111.

65 The importance of repertoire to fitness has been shown for other species; see R. I. Bowman, "A Tribute to the Late Luis Felipe Baptista," in *Nature's Music: The Science of Birdsong*, ed. P. Marler and H. Slabbekoorn (Amsterdam: Elsevier, 2004), 15.

66 Ibid., 33.

67 In 1942 the broadcast was aborted when the sounds of Wellington and Lancaster bombers leaving for a raid were picked up by the microphones. A quick-witted engineer had realized that the live broadcast would give an early warning to the Germans. See "The Remarkable Moment the BBC

Were Forced to Pull Plug on World War II Birdsong Broadcast as Bombers Flew Overhead," *Daily Mail*, January 28, 2012, http://www.dailymail.co.uk/ news/article-2093108/The-remarkable-moment-BBC-forced-pull-plug -World-War-II-birdsong-broadcast-bombers-flew-overhead.html.

68 If I were to pen a letter to Andrew Whitehouse, I would write about puffins. These black-and-white seabirds have bright red bills and are called "bird clowns" because they look so comical. They nest underground and chat with their neighbors through the thin earth walls, growling and purring, making sounds like slowed-down sarcastic laughter.

4: Echoes of the Past

1 The first mention of this phrase comes from the Usenet group talk.politics .mideast, January 8, 1993, when the phrase "A duck can quack but his quack never echoes" appears in the signature of a post.

2 R. Plot, *The Natural History of Oxford-shire, Being an Essay towards the Natural History of England*, 2nd ed. (Oxford: Printed by Leon Lichfield, 1705), 7.

3 *Anechoic* comes from the Greek *an*, meaning "without"; and *echoic*, "relating to an echo."

4 The cave can still be visited today, and by all accounts this is an amazing soundscape to visit. A short piece of avian echolocation can be found in P. Mahler and H. Slabbekoorn, eds., *Nature's Music: The Science of Birdsong* (Amsterdam: Elsevier, 2004), 275.

5 *Encyclopaedia Britannica*, "Marin Mersenne," http://www.britannica.com/ EBchecked/topic/376410/Marin-Mersenne, accessed January 5, 2012.

6 A royal foot is an old measurement of distance. I have taken 1 royal foot to be 0.3287 meter.

7 F. V. Hunt, *Origins in Acoustics: The Science of Sound from Antiquity to the Age of Newton* (New Haven, CT: Yale University Press, 1978), 97. 340 meters per second is the value at 15°C (59°F). The speed of sound depends on temperature.

8 The assumption here is that a quack lasts about 0.19 second and a duck's foot is 5 centimeters (2 inches) long. This would then be a *disyllabic echo*, which is the name for an echo consisting of two syllables, according to the old taxonomy.

9 A slightly simplistic calculation, but probably reasonably accurate, unless there is a temperature inversion that allows the quack to carry farther with

less attenuation. In many tables, the sound level of a rural setting is shown to be 30 decibels and a whisper, 20 decibels.

10 Hunt, *Origins in Acoustics*, 96.

11 Yodeling is more than just a kitsch form of entertainment. It probably originated to enable easy communication in the mountains. The characteristic switches from high to low pitch, as the singer switches from falsetto to a more normal singing voice, make it easier to pick out the yodel after it has traveled a long way and is barely audible.

12 J. McConnachie, *The Rough Guide to the Loire* (London: Rough Guides, 2009), 105.

13 R. Radau, *Wonders of Acoustics* (New York: Scribner, 1870), 82.

14 M. Riffaterre, *Semiotics of Poetry* (Bloomington: Indiana University Press, 1978), 20.

15 The structure drawn by Kircher would need panels spaced at something like 267, 445, 657, and 767 meters (290, 485, 720, and 840 yards) from the listener. It would probably be impossible to shout the initial *"clamore"* loud enough to make the final echo audible, because the panel is too far away. However, by replacing the flat panels drawn by Kircher with concave surfaces, it should be possible to amplify the reflections and overcome this problem.

16 S. B. Thorne and P. Himelstein, "The Role of Suggestion in the Perception of Satanic Messages in Rock-and-Roll Recordings," *Journal of Psychology* 116 (1984): 245–48.

17 I. M. Begg, D. R. Needham, and M. Bookbinder, "Do Backward Messages Unconsciously Affect Listeners? No," *Canadian Journal of Experimental Psychology* 47 (1993): 1–14.

18 *The Simpsons* [television program], season 19, episode 8, first broadcast November 25, 2007, http://movie.subtitlr.com/subtitle/show/190301, accessed November 23, 2011.

19 Radau, *Wonders of Acoustics*, 85–86.

20 P. Doyle, *Echo and Reverb: Fabricating Space in Popular Music, 1900–1960* (Middletown, CT: Wesleyan University Press, 2005), 208.

21 C. Wiley, *The Road Less Travelled: 1,000 Amazing Places off the Tourist Trail* (London: Dorling Kindersley, 2011), 121.

22 R. M. Schafer, *The Soundscape: Our Sonic Environment and the Tuning of the World* (Rochester, VT: Destiny Books, 1994), 220.

23 Luke Jerram, personal communication, October 20, 2011.

24 Whether you find minor scales sad depends on your musical experiences and tastes. If I were from eastern Europe, the sound might not have been so malevolent and spooky.

25 R. Jovanovic, *Perfect Sound Forever* (Boston: Justin, Charles & Co., 2004), 23.

26 Nick Whitaker, acoustic consultant, personal communication, autumn 2011.

27 B. F. G. Katz, O. Delarozière, and P. Luizard, "A Ceiling Case Study Inspired by an Historical Scale Model," *Proceedings of the Institute of Acoustics* 33 (2011): 314–24.

28 A. Lepage, "Le Tribunal de l'Abbaye," *Le Monde Illustré* 19 (1875): 373–76. Translated in Katz, Delarozière, and Luizard, "Ceiling Case Study."

29 D. Shiga, "Telescope Could Focus Light without a Mirror or Lens," *New Scientist*, May 1, 2008, http://www.newscientist.com/article/dn13820-tele scope-could-focus-light-without-a-mirror-or-lens.html?full=true.

30 "Echo Saved Ship from Iceberg," *Day* (London), June 22, 1914.

31 H. H. Windsor, "Echo Sailing in Dangerous Waters," *Popular Mechanics* 47 (May 1927): 794–97.

32 Ibid.

33 H. H. Windsor, "They Steer by Ear," *Popular Mechanics* 76 (December 1941): 34–36, 180.

34 D. Kish, "Echo Vision: The Man Who Sees with Sound," *New Scientist*, no. 2703 (April 11, 2009): 31–33.

35 Tor Halmrast, "More Combs," *Proceedings of the Institute of Acoustics* 22, no. 2 (2011): 75–82.

36 J. A. M. Rojas, J. A. Hermosilla, R. S. Montero, and P. L. L. Espí, "Physical Analysis of Several Organic Signals for Human Echolocation: Oral Vacuum Pulses," *Acta Acustica united with Acustica* 95 (2009): 325–30.

37 L. D. Rosenblum, *See What I'm Saying* (New York: W. W. Norton, 2010).

38 A. T. Jones, "The Echoes at Echo Bridge," *Journal of the Acoustical Society of America* 20 (1948): 706–7.

39 M. Twain, "The Canvasser's Tale," in *The Complete Short Stories of Mark Twain* (Stilwell, KS: Digireads.com, 2008), 90.

40 Hunt (in *Origins in Acoustics*, p. 96) writes that this is the first verse, but Radau (in *Wonders of Acoustics*, p. 93) states that it is the first line. I use the latter because Hunt gives a timing of 32 seconds for the eight repeats, which is far too short for eight repeats of the first verse, but believable for eight repeats of one line.

41 M. Crunelle, "Is There an Acoustical Tradition in Western Architecture?"

(paper presented at the 12th International Congress on Sound and Vibration, Lisbon, Portugal, July 11–14, 2005).

42 I. Lauterbach, "The Gardens of the Milanese Villeggiatura in the Mid-sixteenth Century," in *The Italian Garden: Art, Design and Culture*, ed. J. D. Hunt (Cambridge: Cambridge University Press, 1996), 150. This was also noted by Athanasius Kircher; see L. Tronchin, "The 'Phonurgia Nova' of Athanasius Kircher: The Marvellous Sound World of 17th Century," *Proceedings of Meetings on Acoustics* 4 (2008): 015002.

43 Crunelle, "Is There an Acoustical Tradition?"

44 Marc Crunelle, personal communication, December 3, 2011.

45 Peter Cusack, personal communication, January 7, 2012.

46 Saxophones are usually made from metal. The sound is dominated by the effect of the bore shape, which is conical, and by the way the reed vibration shapes how the notes begin and end. The shape of the mouthpiece is very important, and Charlie Parker would have used his normal mouthpiece on the plastic sax. I play the straight soprano saxophone, which can sound remarkably like the (wooden) oboe because both are conically bored and both use reeds to create sound.

47 "Whistling Echoes from a Drain Pipe," *New Scientist and Science Journal* 51 (July 1, 1971): 6.

48 The explanation is slightly more complex if the hands and ear are not on the center line of the pipe, but the effect is similar; see E. A. Karlow, "Culvert Whistlers: Harmonizing the Wave and Ray Models," *American Journal of Physics* 68 (2000): 531–39.

49 Nico Declercq, personal communication, autumn 2011.

50 D. Tidoni, "A Balloon for Linz" [video], http://vimeo.com/28686368, accessed December 21, 2011.

51 "Remarkable Echoes," in *The Family Magazine* (Cincinnati, OH: J. A. James, 1841), 107.

5: Going round the Bend

1 W. C. Sabine, *Collected Papers on Acoustics* (Cambridge, MA: Harvard University Press, 1922), 257.

2 C. V. Raman, "On Whispering Galleries," *Bulletin of the Indian Association for the Cultivation of Science* 7 (1922): 159–72.

3 E. Boid, *Travels through Sicily and the Lipari Islands, in the Month of December,*

1824, by a Naval Officer (London: T. Flint, 1827), 155–56. W. C. Sabine has interesting discussions on the veracity of the story in his collected papers.

4   T. J. Cox, "Comment on article 'Nico F. Declercq et al.: An Acoustic Diffraction Study of a Specifically Designed Auditorium Having a Corrugated Ceiling: Alvar Aalto's Lecture Room,'" *Acta Acustica united with Acustica* 97 (2011): 909.

5   N. Arnott, *Elements of Physics* (London: Printed for Thomas and George Underwood, 1827), xxix–xxx.

6   D. Zimmerman, *Britain's Shield: Radar and the Defeat of the Luftwaffe* (Stroud, Gloucestershire: Sutton, 2001), 22; J. Ferris, "Fighter Defence before Fighter Command: The Rise of Strategic Air Defence in Great Britain, 1917–1934," *Journal of Military History* 63 (1999): 845–84.

7   "Can Sound Really Travel 200 Miles?" *BBC News*, December 13, 2005, http://news.bbc.co.uk/1/hi/magazine/4521232.stm.

8   "Buncefield Oil Depot Explosion 'May Have Damaged Environment for Decades,' Hears Health and Safety Trial," *Daily Mail*, April 15, 2010, http://www.dailymail.co.uk/news/article-1266217/Buncefield-oil-depot-explosion-damaged-environment-decades.html.

9   R. A. Metkemeijer, "The Acoustics of the Auditorium of the Royal Albert Hall before and after Redevelopment," *Proceedings of the Institute of Acoustics*, 19, no. 3 (2002): 57–66.

10 I first found this exercise described in L. Cremer and H. A. Muller. *Principles and Applications of Room Acoustics*. Translated by T. J. Schultz (London: Applied Science, 1982).

11 "Tests Explain Mystery of 'Whispering Galleries,'" *Popular Science Monthly* 129 (October 1936): 21.

12 "Tourists Fill Washington: Nation's Capital the Mecca of Many Sightseers," *New York Times*, April 16, 1894.

13 "A Hall of Statuary: An Interesting Spot at the Great Capitol," *Lewiston Daily Sun*, December 9, 1893.

14 Cremer and Muller, *Principles and Applications*.

15 There is much more on this design method in my academic text on the subject: T. J. Cox and P. D'Antonio, *Acoustic Absorbers and Diffusers*, 2nd ed. (London: Taylor & Francis, 2009).

16 M. Kington, "Millennium Dome 3, St Peter's Dome 1," *Independent* (London), October 23, 2000.

17 W. Hartmann, H. S. Colburn, and G. Kidd, "Mapparium Acoustics" (lay

language paper presented at the 151st Acoustical Society of America Meeting, Providence, RI, June 5, 2006), http://www.acoustics.org/press/151st/Hartmann.html, accessed February 2011.

18 J. Sánchez-Dehesa, A. Håkansson, F. Cervera, F. Mesegner, B. Manzanares-Martínez, and F. Ramos-Mendieta, "Acoustical Phenomenon in Ancient Totonac's Monument" (lay language paper presented at the 147th Acoustical Society of America Meeting, New York, May 28, 2004), http://www.acoustics.org/press/147th/sanchez.htm, accessed February 2011.

19 R. Godwin, "On a Mission with London's Urban Explorers," *London Evening Standard*, June 15, 2012; A. Craddock, "Underground Ghost Station Explorers Spook the Security Services," *Guardian* (London), February 24, 2012; and B. L. Garrett, "Place Hacking: Explore Everything," *Vimeo*, http://vimeo.com/channels/placehacking, accessed December 29, 2012.

20 It is estimated that there was 55–80 million cubic meters (about 70–105 million cubic yards) of rubble in Berlin to dispose of after the war. The mountain buried a Nazi military training school.

21 Cremer and Muller, *Principles and Applications.*

22 Hartmann, Colburn, and Kidd, "Mapparium Acoustics."

23 Ibid.

24 Barry Marshall, personal communication, May 13, 2011.

25 M. Crunelle, "Is There an Acoustical Tradition in Western Architecture?" http://www.wseas.us/e-library/conferences/skiathos2001/papers/102.pdf, accessed December 29, 2012.

26 G. F. Angas, *A Ramble in Malta and Sicily* (London: Smith, Elder, and Co., Cornhill, 1842), 88.

27 T. S. Hughes, *Travels in Sicily, Greece & Albania, Volume* 1 (London: J. Mawman, 1820), 104–5.

28 A. Bigelow, *Travels in Malta and Sicily: With Sketches of Gibraltar, in MDCCCXXVII* (Boston: Carter, Hendee and Babcock, 1831), 303.

29 J. Verne, *A Journey to the Centre of the Earth* (London: Fantastica, 2013), 125–6.

30 The Temple of Heaven in Beijing includes an "Echo Wall" that is actually a whispering wall.

31 E. C. Everbach and D. Lubman, "Whispering Arches as Intimate Soundscapes," *Journal of the Acoustical Society of America* 127 (2010): 1933.

32 Lord Rayleigh, *Scientific Papers. Volume V* (Cambridge: Cambridge University Press, 1912), 171.

33 Raman, "On Whispering Galleries."

34 S. Hedengren, "Audio Ease Releases Acoustics of Indian Monument Gol Gumbaz, One of the Richest Reverbs in the World," *ProTooler* (blog), September 21, 2007, http://www.protoolerblog.com/2007/09/21/audio-ease -releases-acoustics-of-indian-monument-gol-gumbaz-one-of-the-richest -reverbs-in-the-world.

35 Ibid.

36 "The Missouri Capitol: The Exterior of the Jefferson City Structure Was Built Entirely of Missouri Marble," *Through the Ages Magazine* 1 (1924): 26–32.

37 *A Handbook for Travellers in India, Burma, and Ceylon*, 8th ed. (London: John Murray, 1911).

6: Singing Sands

1 M. L. Hunt and N. M. Vriend, "Booming Sand Dunes," *Annual Review of Earth and Planetary Sciences* 38 (2010): 281–301. There are probably more singing dunes to be found.

2 C. Darwin, *Voyage of the Beagle* (Stilwell, KS: Digireads.com, 2007), 224.

3 L. Giles, "Notes on the District of Tun-Huang," *Journal of the Royal Asiatic Society* 46 (1914): 703–28.

4 M. Polo, *The Travels of Marco Polo* (New York: Cosimo, 2007), 66.

5 S. Dagois-Bohy, S. Ngo, S. C. du Pont, and S. Douady, "Laboratory Singing Sand Avalanches," *Ultrasonics* 50 (2010): 127–32.

6 Data from "World's Largest Waterfalls by Average Volume," *World Waterfall Database*, http://www.worldwaterfalldatabase.com/largest-waterfalls/volume, accessed December 27, 2012.

7 J. Knelman, "Did He or Didn't He? The Canadian Accused of Inventing CIA Torture," *Globe and Mail* (Canada), November 17, 2007.

8 J. Muir, *John Muir: The Eight Wilderness Discovery Books* (Seattle, WA: Diadem, 1992), 623.

9 G. R. Watts, R. J. Pheasant, K. V. Horoshenkov, and L. Ragonesi, "Measurement and Subjective Assessment of Water Generated Sounds," *Acta Acustica united with Acustica* 95: 1032–39 (2009); L. Galbrun and T. T. Ali, "Perceptual Assessment of Water Sounds for Road Traffic Noise Masking," in *Proceedings of the Acoustics 2012 Nantes Conference, 23–27 April 2012, Nantes, France*, 2153–2158.

10 Lee Patterson, personal communication, May 25, 2012.

11 The bubbles have quite a tight range of sizes, from 1 to 3 millimeters (about

a twelfth of an inch) in radius, resulting in a narrow range of frequencies, from 1,000 to 3,000 hertz. It's possible that the bubbles come from a gas produced by insects, but given the fluctuations of the bubble production with light levels that Lee described to me, the source would seem more likely to be photosynthesis from very lively pondweed.

12 L. Rohter, "Far from the Ocean, Surfers Ride Brazil's Endless Wave," *New York Times*, March 22, 2004.

13 For much more on tidal bores, see the excellent book by G. Pretor-Pinney *Wavewatcher's Companion* (London: Bloomsbury, 2010), from which some of the facts in this section are taken.

14 Z. Dai and C. Zhou, "The Qiantang Bore," *International Journal of Sediment Research* 1 (1987): 21–26.

15 W. U. Moore, "The Bore of the Tsien-Tang-Kiang," *Proceedings of the Institution of Civil Engineers* 99 (1890): 297–304.

16 H. Chanson, "The Rumble Sound Generated by a Tidal Bore Event in the Baie du Mont Saint Michel," *Journal of the Acoustical Society of America* 125 (2009): 3561–68.

17 Terje Isungset, comment to the audience at a concert at the Royal Northern College of Music, Manchester, England, November 7, 2011.

18 L. R. Taylor, M. G. Prasad, and R. B. Bhat, "Acoustical Characteristics of a Conch Shell Trumpet," *Journal of the Acoustical Society of America* 95 (1994): 2912.

19 P. Wyse, "The Iceman Bloweth," *Guardian* (London), December 3, 2008.

20 This reminds me of a terrible hi-fi magazine article I once read claiming that the material of the shelf where a CD player rested significantly changed the sound—wooden shelves producing a warm sound, glass shelves producing a clearer sound!

21 B. L. Giordano and S. McAdams, "Material Identification of Real Impact Sounds: Effects of Size Variation in Steel, Glass, Wood, and Plexiglass Plates," *Journal of the Acoustical Society of America* 119 (2006): 1171–81.

22 O. Chernets and J. R. Fricke, "Estimation of Arctic Ice Thickness from Ambient Noise," *Journal of the Acoustical Society of America* 96 (1994): 3232–33.

23 Peter Cusack, personal communication, January 7, 2012.

24 Chris Watson, personal communication, November 15, 2011.

25 R. van der Spuy, *AdvancED Game Design with Flash* (New York: friendsofED, 2010), 462.

26 G. Lundmark, "Skating on Thin Ice—and the Acoustics of Infinite Plates"

(paper presented at Internoise 2001, the 2001 International Congress and Exhibition on Noise Control Engineering, The Hague, Netherlands, August 27–30, 2001).

27   S. Dagois-Bohy, S. Courrech du Pont, and S. Douady, "Singing-Sand Avalanches without Dunes," *Geophysical Research Letters* 39 (2012): L20310.

28   E. R. Yarham, "Mystery of Singing Sands," *Natural History* 56 (1947): 324–25.

29   C. Grant, *Rock Art of the American Indian* (Dillon, CO: VistaBooks, 1992).

30   "Sound Effects: Castle Thunder," *Hollywood Lost and Found*, http://www .hollywoodlostandfound.net/sound/castlethunder.html, accessed December 31, 2012.

31   Tim Gedemer, personal communication, June 24, 2012.

32   Heat is the normal explanation for why the shock wave forms; for example, see F. Blanco, P. La Rocca, C. Petta, and F. Riggi, "Modelling Digital Thunder," *European Journal of Physics* 30 (2009): 139–45. But another suggestion is that the energy for the sound waves comes from breaking chemical bonds; see P. Graneau, "The Cause of Thunder," *Journal of Physics D: Applied Physics* 22 (1989): 1083–94.

33   H. S. Ribner and D. Roy, "Acoustics of Thunder: A Quasilinear Model for Tortuous Lightning," *Journal of the Acoustical Society of America* 72 (1982): 1911–25.

34   D. P. Hill, "What Is That Mysterious Booming Sound?" *Seismology Research Letters* 82 (2011): 619–22.

35   D. Ramde and C. Antlfinger, "Wis. Town Longs for Relief from Mysterious Booms," *Associated Press*, March 21, 2012.

36   J. Van Berkel, "Data Point to Earthquakes Causing Mysterious Wis. Booms," *USA Today*, March 22, 2012.

37   C. Davidson, "Earthquake Sounds," *Bulletin of the Seismological Society of America* 28 (1938): 147–61.

38   L. Bogustawski, "Jets Make Sonic Boom in False Alarm," *Guardian* (London), April 12, 2012.

39   For more stories about Krakatoa, see S. Winchester, *Krakatoa: The Day the World Exploded: August 27, 1883* (New York: Harper-Collins, 2005). The various contemporary quotes are extracted from this book.

40   Although no instrument existed to give a precise value of loudness, eyewitness reports have led to a commonly quoted figure of 180 decibels at 160 kilometers (100 miles) from the volcano. But I have been unable to find the origin of this estimation.

41 D. Leffman, *The Rough Guide to Iceland* (London: Rough Guides, 2004), 277.

42 Tim Leighton, personal communication, March 1, 2012.

43 The speed of the water ejection can exceed the speed of sound, leading to small sonic booms that make thudding sounds; see T. S. Bryan, *The Geysers of Yellowstone*, 4th ed. (Boulder: University Press of Colorado, 2008), 5–6.

44 Darwin, *Voyage of the Beagle*.

45 T. Hardy, *Under the Greenwood Tree* (Stilwell, KS: Digireads.com, 2007), 7.

46 O. Fégeant, "Wind-Induced Vegetation Noise. Part I: A Prediction Model," *Acustica united with Acta Acustica* 85 (1999): 228–40.

47 I was told this by garden designer Paul Hervey Brookes during an interview for BBC Radio 4 conducted May 20, 2011.

48 C. M. Ward, "Papers of Mel (Charles Melbourne) Ward," AMS 358, Box 3, Notebook 31 (Sydney: Australian Museum, 1939). The two quotes are from C. A. Pocock, "Romancing the Reef: History, Heritage and the Hyper-real" (PhD dissertation, James Cook University, Australia, 2003).

49 Y. Qureshi, "Tower Blows the Whistle on Corrie," *Manchester Evening News*, May 24, 2006.

50 "Beetham Tower Howls Again after Another Windy Night in Manchester," *Manchester Evening News*, January 5, 2012. Before, the tower used to make noise at about 50 kilometers (30 miles) per hour; now it sings only when the wind speed exceeds 110 kilometers (70 miles) per hour.

51 G. Sargent, "I'm Sorry about the Beetham Tower Howl, Says Architect Ian Simpson," *Manchester Evening News*, January 6, 2012. For other ways that buildings make noise, see M. Hamer, "Buildings That Whistle in the Wind," *New Scientist*, no. 2563 (August 4, 2006): 34–36.

52 A. R. Gold, "Ear-Piercing Skyscraper Whistles Up a Gag Order," *New York Times*, April 13, 1991.

53 Simon Jackson, of the acoustic-consulting firm Arup, tweeted, "Quick sound level measurement at Beetham Tower - 78dBLaeq,1s main freq in 250Hz 3rd/oct band" (@stjackson, January 3, 2012).

54 Nathalie Vriend, personal communication, February 17, 2012.

55 Curzon of Kedleston, Marquess, *Tales of Travel* (New York: George H. Doran, 1923), 261–339.

7: The Quietest Places in the World

1 C. D. Geisler, *From Sound to Synapse: Physiology of the Mammalian Ear* (New York: Oxford University Press, 1998), 194.

2   J. J. Eggermont and L. E. Roberts, "The Neuroscience of Tinnitus," *Trends in Neurosciences* 27 (2004): 676–82.

3   R. Schaette and D. McAlpine, "Tinnitus with a Normal Audiogram: Physiological Evidence for Hidden Hearing Loss and Computational Model," *Journal of Neuroscience* 31 (2011): 13452–57.

4   C. Watson, "No Silence Please," *Inside Music* (BBC blog), December 2006, http://www.bbc.co.uk/music/insidemusic/nosilenceplease, accessed June 1, 2009.

5   To be precise, it is –9.4 dBA. The *A* means this is an A-weighted decibel, where corrections are applied to allow for the fact that the ear is less sensitive at low frequencies.

6   It is still useful to make the halls very quiet, because audience noise is reduced in quieter auditoriums. C.-H. Jeong, M. Pierre, J. Brunskog, and C. M. Petersen, "Audience Noise in Concert Halls during Musical Performances," *Journal of the Acoustical Society of America* 131 (2012): 2753–61.

7   M. Botha, "Several Futures of Silence: A Conversation with Stuart Sim on Noise and Silence," *Kaleidoscope* 1, no. 1 (2007), https://www.dur.ac.uk/kaleidoscope/issues/i1v1/sim_1_1. Stuart Sim authored *A Manifesto for Silence* (Edinburgh: Edinburgh University Press, 2007).

8   J. A. Grahn, "Neural Mechanisms of Rhythm Perception: Current Findings and Future Perspectives," *Topics in Cognitive Science* 4 (2012): 585–606.

9   D. Levitin, *This Is Your Brain on Music* (London: Atlantic, 2006).

10   Charles Deenen, personal communication, June 24, 2012.

11   See either K. Young, "Noisy ISS May Have Damaged Astronauts," *New Scientist*, June 2006; or C. A. Roller and J. B. Clark, "Short-Duration Space Flight and Hearing Loss," *Otolaryngology—Head and Neck Surgery* 129 (2003): 98–106.

12   Young, "Noisy ISS."

13   T. G. Leighton and A. Petculescu, "The Sound of Music and Voices in Space Part 2: Modeling and Simulation," *Acoustics Today* 5 (2009): 17–26.

14   A. Moorhouse, *Environmental Noise and Health in the UK* (Oxfordshire, UK: Health Protection Agency, 2010).

15   J. Voisin, A. Bidet-Caulet, O. Bertrand, and P. Fonlupt, "Listening in Silence Activates Auditory Areas: A Functional Magnetic Resonance Imaging Study," *Journal of Neuroscience* 26 (2006): 273–78.

16   D. Van Dierendonck and J. T. Nijenhuis, "Flotation Restricted Environmental Stimulation Therapy (REST) as a Stress-Management Tool: A Meta-analysis," *Psychology and Health* 20 (2005): 405–12.

17 H. Samuel, "French Told Not to Complain about Rural Noise," *Daily Telegraph* (London), August 22, 2007.

18 C. Ray, "Soundscapes and the Rural: A Conceptual Review from a British Perspective," Centre for Rural Economy Discussion Paper no. 5, February 2006, http://www.ncl.ac.uk/cre/publish/discussionpapers/pdfs/dp5.pdf.

19 Publisher's note from G. Hempton and J. Grossmann, *One Square Inch of Silence: One Man's Search for Natural Silence in a Noisy World* (New York: Free Press, 2009).

20 "What Is One Square Inch?" *One Square Inch: A Sanctuary for Silence at Olympic National Park*, http://onesquareinch.org/about, accessed July 5, 2013.

21 R. M. Schafer, *The Tuning of the World* (Toronto: McClelland and Stewart, 1977).

22 US National Park Service, *Management Policies* (Washington, DC: US Department of the Interior, 2006), 56.

23 The research underpinning the campaign is given in great detail in S. Jackson, D. Fuller, H. Dunsford, R. Mowbray, S. Hext, R. MacFarlane, and C. Haggett, *Tranquillity Mapping: Developing a Robust Methodology for Planning Support* (Centre for Environmental and Spatial Analysis, Northumbria University, 2008).

24 B. L. Mace, P. A. Bell, and R. J. Loomis, "Aesthetic, Affective and Cognitive Effects of Noise on Natural Landscape Assessment," *Social & Natural Resources* 12 (1999): 225–42.

25 M. D. Hunter, S. B. Eickhoff, R. J. Pheasant, M. J. Douglas, G. R. Watts, T. F. Farrow, D. Hyland, et al., "The State of Tranquility: Subjective Perception Is Shaped by Contextual Modulation of Auditory Connectivity," *NeuroImage* 53 (2010): 611–18.

26 S. Maitland, *A Book of Silence* (London: Granta, 2008).

27 S. Arkette, "Sounds like City," *Theory, Culture & Society* 21 (2004): 159–68.

28 Jackson et al., *Tranquillity Mapping*.

29 Based on findings in the UK. C. J. Skinner and C. J. Grimwood, *The UK National Noise Incidence Study* 2000/2001, vol. 1, *Noise Levels* (London: Department for Environment, Food & Rural Affairs, 2001).

30 "Directive 2002/49/EC of the European Parliament and of the Council of 25 June 2002 Relating to the Assessment and Management of Environmental Noise," *Official Journal of the European Communities* L189/12, July 18, 2002, http://eur-lex.europa.eu/LexUriServ/LexUriServ.do?uri=OJ:L:2002:189:0012:0025:EN:PDF.

31 For the acoustic geeks, the first is L_{den} = 55 dB [*Research into Quiet Areas:*

Recommendations for Identification (London: Department for Environment, Food & Rural Affairs, 2006)]; and the second, $L_{A,eq}$ = 42 dB [R. Pheasant, K. Horoshenkov, G. Watts, and B. Barrett, "The Acoustic and Visual Factors Influencing the Construction of Tranquil Space in Urban and Rural Environments Tranquil Spaces-Quiet Places?" *Journal of the Acoustical Society of America* 123 (2008): 1446–57].

32   Hildegard Westerkamp, personal communication, April 19, 2009.

33   Other studies have come up with sets of terms similar to *vibrancy* and *pleasantness*. W. J. Davies and J. E. Murphy, "Reproducibility of Soundscape Dimensions" (paper presented at InterNoise 2012, New York, August 19–22, 2012).

34   Stuart Bradley, personal communication, April 2009.

35   W. Hasenkamp and L. W. Barsalou, "Effects of Meditation Experience on Functional Connectivity of Distributed Brain Networks," *Frontiers in Human Neuroscience* 6 (2012): 38.

36   Wendy explained to me that while older studies focused on individual brain regions, newer brain-imaging techniques have shown that no region of the brain is solely responsible for any given function, and that functions are distributed across multiple brain regions called networks.

37   K. A. MacLean, E. Ferrer, S. R. Aichele, D. A. Bridwell, A. P. Zanesco, T. L. Jacobs, B. G. King, et al., "Intensive Meditation Training Improves Perceptual Discrimination and Sustained Attention," *Psychological Science* 21 (2010): 829–839.

8: Placing Sound

1   For an analysis of randomness in musical rhythms over four centuries, see D. J. Levitin, P. Chordia, and V. Menon, "Musical Rhythm Spectra from Bach to Joplin Obey a $1/f$ Power Law," *Proceedings of the National Academy of Sciences of the USA* 109 (2012): 3716–20.

2   "New Organ Will Be Played by the Sea," *Lancashire Telegraph*, June 14, 2002.

3   S.-H. Kima, C.-W. Lee, and J.-M. Lee, "Beat Characteristics and Beat Maps of the King Seong-deok Divine Bell," *Journal of Sound and Vibration* 281 (2005): 21–44.

4   Some musical instruments produce notes that lack the fundamental, but in that case the brain fills in the missing information.

5   The frequency is proportional to the wind speed divided by the wire thickness.

6   A. Hickling, "Blowing in the Wind: Pierre Sauvageot's Harmonic Fields," *Guardian* (London), June 2, 2011, http://www.guardian.co.uk/music/2011/jun/02/harmonic-fields-pierre-sauvageot.

7   M. Kamo and Y. Iwasa, "Evolution of Preference for Consonances as a By-product," *Evolutionary Ecology Research* 2 (2000): 375–83.

8   See N. Bannan, ed., *Music, Language, and Human Evolution* (Oxford: Oxford University Press, 2012).

9   A. Corbin, "Identity Bells and the Nineteenth Century French Village," in M. M. Smith, *Hearing History* (Athens: University of Georgia Press, 2004), 184–200.

10 My visit to St. James' Church near Manchester took place on September 10, 2011.

11 The "brown bread" line was taken from a *Daily Mail* headline, June 26, 2012.

12 T. J. Cox, "Acoustic Iridescence," *Journal of the Acoustical Society of America* 129 (2011): 1165–72.

13 P. Ball, "Sculpted Sound," *New Scientist*, no. 2335 (March 23, 2002): 32.

14 A. Climente, D. Torrent, and J. Sánchez-Dehesa, "Omnidirectional Broadband Acoustic Absorber Based on Metamaterials," *Applied Physics Letters* 100 (2012): 144103. My colleague Olga Umnova built a giant one of these to investigate how it can protect you from a blast.

15 Francis Crow, personal communication, November 7, 2012.

16 The sound propagation was probably similar to that of Echo Bridge, Massachusetts, which I discussed in Chapter 4.

17 The quotes in this discussion, with some additional minor changes by Davide Tidoni, were first used as notes to the exhibition "Bang! Being the Building," which was at the Barbican, London, in 2012.

18 "Somerset Church Bell to Ring Again After Agreement Reached," *BBC News*, December 2, 2012, http://www.bbc.co.uk/news/uk-england-somerset-20572854.

19 Angus Carlyle, personal communication, October 19, 2012.

20 Tony Gibbs, personal communication, October 23, 2012.

21 In North Wales there is a project called Bangor Sound City, which has a grand vision for a permanent sound art park, a sonic complement to sculpture parks. The project has staged a series of temporary interventions to examine people's attitudes to public sound art. They found that people take sound exhibits at face value; there was no sense that they hanker after more conventional sculpture or a painting.

22 If you prefer a more eco-friendly version of a musical road, in Rotterdam

there is a floor sculpture in Schouwburgplein that plays with footstep sounds by having different types of paving.

23   David Simmons-Duffin wrote his blog post on December 23, 2008.

24   The distances have been rounded; they were 12.3 centimeters (4.8 inches) for the lowest note and 8.0 centimeters (3.1 inches) for the highest.

25   In musical terms, the last note was between a fifth and a sixth above the first note.

26   There is at least one that does not use octaves, and some omit fifths. P. Ball, "Harmonious Minds: The Hunt for Universal Music," *New Scientist*, no. 2759 (May 10, 2010): 30–33.

27   M. Hamer, "Music Special: Flexible Scales and Immutable Octaves," *New Scientist*, no. 2644 (February 23, 2008): 32–34.

28   "Lone Ranger Road Music Heads into the Sunset," *CBC News*, September 21, 2008, http://www.cbc.ca/news/arts/music/story/2008/09/21/road-lone ranger-theme.html. The quote is originally from "Honda Makes GROOVY Music," *Daily Breeze*, September 20, 2008, http://www.dailybreeze.com/ci _10514483.

9: Future Wonders

1   The food industry also manipulates product sound, to ensure that biscuits and snacks have the right crispiness and crunchiness. For crispiness, recipes need to create a brittle internal scaffold within the food that shatters in your mouth when you first bite.

2   P. Nyeste and M. S. Wogalter, "On Adding Sound to Quiet Vehicles," in *Proceedings of the Human Factors and Ergonomics Society 52nd Annual Meeting—2008* (Human Factors and Ergonomics Society, 2008), 1747–50.

3   "Feedback: Cars That Go Clippetty Clop," *New Scientist*, no. 2823 (July 29, 2011).

4   *The Museum of Curiosity*, series 2, episode 1, BBC Radio 4, broadcast May 4, 2009.

5   "99% Invisible-15- Sounds of the Artificial World," *PodBean*, February 11, 2011, http://podbean.99percentinvisible.org/2011/02/11/99-invisible-15-sounds -of-the-artificial-world.

6   S. Koelsch, "Effects of Unexpected Chords and of Performer's Expression on Brain Responses and Electrodermal Activity," *PLoS One* 3, no. 7 (2008): e2631.

7   The reproduction often suffers from the fact that the sound appears to be inside the head. Researchers are currently investigating how this problem might be overcome for all listeners. The story of the opera house comes from A. Farina and R. Ayalon, "Recording Concert Hall Acoustics for Posterity" (paper presented at the 24th International Conference: Multichannel Audio—The New Reality, June 26–28, 2003).

Index

Page numbers in *italics* refer to illustrations.
Page numbers beginning with 281 refer to notes.